Erika Di Marino

Ectomycorrhizal community in beech coppices of different age

Erika Di Marino

Ectomycorrhizal community in beech coppices of different age

Dissertation Ludwig-Maximilians Universität München Fakultät für Biologie, Phd Thesis Università degli Studi di Padova Scuola di Dottorato di ricerca Territorio Ambiente Risorse e salute

Südwestdeutscher Verlag für Hochschulschriften

Impressum/Imprint (nur für Deutschland/ only for Germany)
Bibliografische Information der Deutschen Nationalbibliothek: Die Deutsche Nationalbibliothek verzeichnet diese Publikation in der Deutschen Nationalbibliografie; detaillierte bibliografische Daten sind im Internet über http://dnb.d-nb.de abrufbar.

Alle in diesem Buch genannten Marken und Produktnamen unterliegen warenzeichen-, marken- oder patentrechtlichem Schutz bzw. sind Warenzeichen oder eingetragene Warenzeichen der jeweiligen Inhaber. Die Wiedergabe von Marken, Produktnamen, Gebrauchsnamen, Handelsnamen, Warenbezeichnungen u.s.w. in diesem Werk berechtigt auch ohne besondere Kennzeichnung nicht zu der Annahme, dass solche Namen im Sinne der Warenzeichen- und Markenschutzgesetzgebung als frei zu betrachten wären und daher von jedermann benutzt werden dürften.

Verlag: Südwestdeutscher Verlag für Hochschulschriften Aktiengesellschaft & Co. KG
Dudweiler Landstr. 99, 66123 Saarbrücken, Deutschland
Telefon +49 681 37 20 271-1, Telefax +49 681 37 20 271-0
Email: info@svh-verlag.de
Zugl.: Muenchen, LMU, Diss. 2009

Herstellung in Deutschland:
Schaltungsdienst Lange o.H.G., Berlin
Books on Demand GmbH, Norderstedt
Reha GmbH, Saarbrücken
Amazon Distribution GmbH, Leipzig
ISBN: 978-3-8381-0802-5

Imprint (only for USA, GB)
Bibliographic information published by the Deutsche Nationalbibliothek: The Deutsche Nationalbibliothek lists this publication in the Deutsche Nationalbibliografie; detailed bibliographic data are available in the Internet at http://dnb.d-nb.de.

Any brand names and product names mentioned in this book are subject to trademark, brand or patent protection and are trademarks or registered trademarks of their respective holders. The use of brand names, product names, common names, trade names, product descriptions etc. even without a particular marking in this works is in no way to be construed to mean that such names may be regarded as unrestricted in respect of trademark and brand protection legislation and could thus be used by anyone.

Publisher: Südwestdeutscher Verlag für Hochschulschriften Aktiengesellschaft & Co. KG
Dudweiler Landstr. 99, 66123 Saarbrücken, Germany
Phone +49 681 37 20 271-1, Fax +49 681 37 20 271-0
Email: info@svh-verlag.de

Printed in the U.S.A.
Printed in the U.K. by (see last page)
ISBN: 978-3-8381-0802-5

Copyright © 2010 by the author and Südwestdeutscher Verlag für Hochschulschriften Aktiengesellschaft & Co. KG and licensors
All rights reserved. Saarbrücken 2010

UNIVERSITÀ DEGLI STUDI DI PADOVA
FACOLTÀ DI AGRARIA

DIPARTIMENTO TERRITORIO E SISTEMI AGRO-FORESTALI

LUDWIG-MAXIMILIANS-UNIVERSITÄT MÜNCHEN
FAKULTÄT FÜR BIOLOGIE

DEPARTMENT BIOLOGIE

SCUOLA DI DOTTORATO DI RICERCA IN : TERRITORIO, AMBIENTE, RISORSE E SALUTE

INDIRIZZO: Ecologia

CICLO XX

THE ECTOMYCORRHIZAL COMMUNITY STRUCTURE IN BEECH COPPICES OF DIFFERENT AGE

Direttore della Scuola : Ch.mo Prof. Vasco Boatto
Supervisore : Ch.mo Prof. Lucio Montecchio
Supervisore : Ch.mo Prof. Reinhard Agerer

Dottoranda : Erika Di Marino

DATA CONSEGNA TESI
31 gennaio 2008

CONTENTS

Chapter 1: Introduction   3

Chapter 2: The composition of the ectomycorrhizal community in beech coppices of different age   49

Chapter 3: (Papee I) Hygrophorus penarius on beech: between mutualism and parasitism?   88

Chapter 4: (Draft) The ectomycorrhizal community structure in beech coppices differing in age and stand features – short communication: Distribution of ectomycorrhizae in beech coppices regarding exploration types and hydrophobicity-hydrophily features   112

Chapter 5: (Description) *"Fagirhiza entolomoides"* + *Fagus sylvatica* (L.)   155

Chapter 6: (Description) *"Fagirhiza bissoporoides"* + *Fagus sylvatica* (L.)   166

Chapter 7: (Description*) "Fagirhiza stellata"* + *Fagus sylvatica (L.)*   180

Chapter 8: (Paper II*) Sistotrema* is a genus with ectomycorrhizal species – confirmation of what sequences studies already suggested   193

Chapter 9: (Paper III) The ectomycorrhzae of *Pseudotomentella humicola* on *Picea abies*   210

Chapter 10: Conclusion   228

Abstract

Acknowledgments

The experimental works described in this thesis are part of scientific paper pubblicated, submitted or to be submitted to international journal.

Paper I submitted the 29th Jenuary to *Mycorrhiza*

Paper II sumbmitted the 11th Jenuary to *Mycological Research*

Paper III received the 6th July 2006 accepted in revised form the 31th July 2006 *Nova Edwigia* 84 (3-4): 429-440.

CHAPTER 1

Introduction

1. Ectomycorrhizal symbiosis

The German term *"Symbiotismus"* (symbiosis) was probably first used by Frank (1877) as a neutral term that did not imply parasitism, but was based simply on the regular coexistence of dissimilar organisms, such as is observed in lichens (Smith & Read, 1997). De Bary (1887) used it to identify the common life of parasite and host as well as of associations in which the organisms apparently help each other. Since then the meaning of the terms symbiosis and parasite have changed, with symbiosis being used more and more for mutually beneficial associations between dissimilar organisms, and parasite and parasitism being almost synonymous with *pathogen* and *pathogenesis* (Smith & Read, 1997). De Bary also pointed out that there is every conceivable gradation between the parasite that quickly destroys its victim and those that "further and support" their partners, and in recent years researchers have come back to this view.

Although generally the mycorrhizal symbioses are considered mutualistic, due to the benefits for both partners, a better description probably is that individual plant and fungal symbionts are placed somewhere in the mutualistic-parasitic continuum, depending on their developmental state, the specific genotype combinations and the environmental conditions (Johnson *et al.*, 1997; Egger & Hibbet, 2004). This thesis is based on the word "symbiosis", as defined by de Bary. An example of a modified relationship between the plant and fungi, is a mycorrhiza that becomes a specimen of parasitism, where the mantle is present but the Hartig net is lacking is also reported (Paper I-chapter 3).

The EM (ectomycorrhizal) symbiosis is typically formed between the terminal feeder roots of woody perennial plant species and a range of soil fungi (Smith & Read, 1997). The fungi exchange soil-derived nutrients for carbohydrates from the host plant. Nutrient uptake into the host is enhanced both as a consequence of the physical geometry of the fungal mycelium and by the ability of the fungi to mobilise N and P from organic substrates through the action of secreted catabolic enzymes (Leake & Read, 1997).

Within the root, the fungus ramifies between the outer cells forming a complex structure called the Hartig net (fig. 1), which provides a large surface area of contact between the fungus and the host, allowing an efficient transfer of metabolites. External to the root, a multi-layered, hyphal structure, called the mantle or sheath, develops (Taylor & Alexander, 2005).

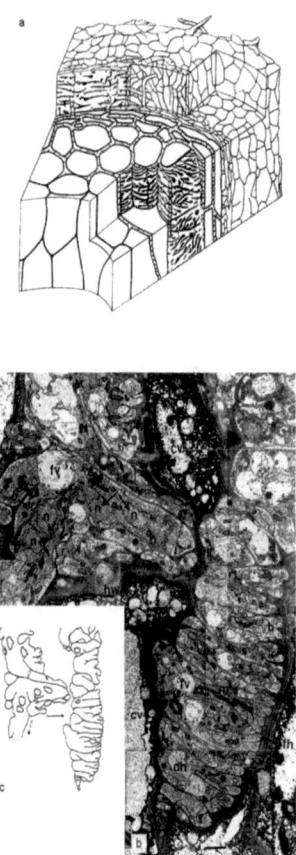

Fig. 1: Development of the Hartig net (modified from Smith & Read, 1997). (a) Block diagram showing typical structure of the Hartig net in different sectional aspects and of a pseudoparenchymatous mantle. The main growth direction of the hyphae in the Hartig net is transverse to the root axis. (b) Transmission electron microscopy of a mycorrhiza formed between *Picea abies* and *Amanita muscaria*. Ultrathin section through the intercellular space and several cortical cells showing fully developed, mature Hartig net. Extensive branching leads to the formation of narrower and narrower hyphae (fh). Numerous mitochondria (m) and nuclei (n) can be seen. The presence of two dikaryons (arrowed) indicates the coenocytic nature of the tissue. (FV) fungal vavuole; (n) nucleus; (cv) epidermal cell

vacuole, (hw) host wall. Bar, 2 µm. (c) Outline of the hyphae in (b). Main growth of the hyphae is in the direction of the full arrow. Dolipore septum and dikaryons are marked. From Kottke and Oberwinkler (1987).

Agerer (1987-2002) has recognised two main types of hyphal development within EM mantles: *pseudoparenchymatous* – densely packed, highly differentiated hyphal elements, and *plectenchymatous* - loosely interwoven hyphae, where their linear nature is still evident. The hyphal arrangement within the mantle, particularly when seen in plain view, has been used by Agerer and coworkers to characterise the mantles formed by individual species as an aid to identification (Agerer 1987-2002; Agerer *et al.*, 1996 – 2004).

Several EM species form mantles that are hydrophobic (e.g. species belong to *Cortinarius* genus, Agerer, 1987-2002), implying that there is little direct exchange of solutes (uptake or exudation) with the soil solution (Taylor & Alexander, 2005). These species, possessing water repellence properties, seem to prefer highly areated soil in the conifer forest soils (Unestam, 1991). Despite this behaviour, the ecological strategy of the hydrophilic fungi is not very clear (Unestam, 1991; Unestam & Stenström,1989; Stenström 1991).

These hydrophilic mantles (e.g. many *Lactarius* species) appear to be a close control over the movement and the exchange of material through the mantle (Ashford *et al.*, 1988), and are most likely responsible for the uptake of water and nutrients (Cairney & Burke, 1996).

EM fungi probably control the interface between the soil environment and the host plant. While the mantles may control the fluxes into and out of the root, the mycelium extending out from the mantle surface in the surrounding soil (the extraradical or extramatrical mycelium) is considered to be the primary site for nutrient and water uptake (Taylor & Alexander, 2005).

The extramatrical mycelia produced by EM fungi varies from a small number of hyphae growing out a few mm (e.g. *Russula* spp.) to highly developed, extensive mycelial systems (e.g. *Suillus* spp., *Cortinarius* spp.) that occupy large volumes of soil surrounding the colonised root tips (Agerer 1987-2002). The extension and the structure of this extramatrical mycelium is thought to be different among EM fungal taxa (Agerer 2001). In this context the purpose to classify the EM fungal species with the "exploration types" according to Agerer (2001) interpreting the anatomical features like ecological strategies to colonise the soil, becomes more and more important to understand the role of these organisms, as key elements of forest nutrient cycles and a strong diversity of forest ecosystem processes (Read *et al.*, 2004). The mycelium formed hydrophilic structures seems to have substrate particles glued to their surface (Raidl 1997) and the hyphae are thicker than the other most distant ones, which have relatively hydrophobic proximal parts (Unestam & Stun, 1995).

This hypothesis was confirmed by a recent study on the production of oxidases of ectomycorrhizal fungi (Agerer *et al.*, 2000) using fruitbodies. An evident correlation between fungal relationship, production of phenoloxidases and exploration type fo their ectomycorrhizae was found, because all *Lactarius* and *Russula* species revealed a higher ability to produce extracellular phenoloxidases. In contrast to almost all members of the order *Boletales*, which lack this feature. The authors correlated these results with the exploration type of their mycorrhizae. Most species of the genus *Russula* and *Lactarius* belong to the "contact exploration type", only some *Lactarius* species the "medium-distance smooth exploration type" (Agerer 2001). The profitable exploration of the surrounding substrate is assured by the ectomycorrhizae of both genera thanks to their hydrophilic behaviour.

The typical ability to degrade lignin of these exploration types, which should increase access to nitrogen complexed to phenolic substances (Kuiters 1990) and could, therefore, support nutrient acquisition when squeezed between organic substrates (Agerer 2001). For these reasons in beech forests the EM belonging to the *Lactarius*, *Russula* and *Laccaria* genera are probably widespread in the upper soil, like a "sandwich" in the superficial thick layers fo leaves and other organic matter (Brand 1991).

The *Boletales*, however, are all known to form ectomycorrhizae of the long-distance exploration type (Agerer 1999; with the exception of *Gomphidiaceae*) and are mostly hydrophobic at their proximal parts. Nutrient acquisition appears to be limited to the very distant hydrophilic substrate adhesion hyphae (Raidl 1997; Unestam & Sun, 1995). The lack of lignin degradation ability is compensated here by a larger surface area and a greater range of spread (Raidl 1997). *Laccaria* species, possibly ascribable to the contact, medium or short-distance exploration type and generally hydrophilic, consistently produce extracellular phenoloxidases (Agerer 2001). However the capacity to produce extracellular phenoloxidases was not generally related to the type of exploration. The genus *Dermocybe*, for example, with its medium-distance fringe exploration type, lacks phenoloxidases. Species- and strain-specific differences were apparent in other genera (Agerer *et al.*, 2000).

The crucial importance of this extramatrical mycelium in nutrient uptake has been emphasized in recent years, and in particular the role of the symbiosis in the facilitation of capture of nitrogen (N) and phosphorus (P) in ionic form (Read & Perez-Moreno, 2003). In addition several recent investigations have utilized molecular markers to localise the mycelium of EM fungal species in different soil layers and substrates (Dickie *et al.*, 2003; Guidot *et al.*, 2003; Landeweert *et al.*, 2003; Koide *et al.*, 2005).

2. The phytobionts

Around 8000 spp., about 3%, of seed plants form EM (Meyer 1973; Smith & Read, 1997) but this minority of plant species is of enormous ecological and economic importance, because they are the dominant components of forest and woodland ecosystems over much of the earth's surface (Taylor & Alexander, 2005). The great majority of EM plants are woody perennials (Fitter & Moyersoen 1996), but also some sedges (*Kobresia* spp.), and herbaceous *Polygonum* spp. form ectomycorrhizas (Massicotte *et al.*, 1998). The forest dominants of the temperate and boreal zone (*Fagaceae, Betulaceae, Salicaceae, Pinaceae*) are habitually ectomycorrhizal under natural conditions, and the EM habit shows particular structures like adaptations for nutrient capture in temperate and boreal forests (Read & Perez-Moreno, 2003). Furthermore the EM occurrence elsewhere is patchy (Taylor & Alexander, 2005). However, much of the rest of the land surface also supports vegetation with a strong EM component: arctic and alpine habitats in the northern hemisphere are characterised by dwarf shrub communities of *Dryas* and *Salix* spp. support EM communities and the winter-wet ecosystems of the Mediterranean basin and California have a strong EM/arbutoid mycorrhizal component (*Pinus, Cistus, Arbutus, Arctostaphylos*; Taylor & Alexander, 2005). In the tropics the occurrence and importance of EM host species has been most consistently underestimated instead (Taylor & Alexander, 2005). In these ecosystems the family *Dipterocarpaceae* with more of 500 spp., all members that form ectomycorrhizas have a key role for the potential spreading of the mycobionts. The range of dipterocarps extends from East Africa and Madagascar, through India, Bangladesh and Sri Lanka, to South-Est Asia, from South China in the north, to Papua New Guinea in the south and there is one genus (*Pakaraimea*) in South America. They also dominate the canopy trees and the understorey in South Est Asian lowland and highland rain forest, dry monsoonal forests of North India, Burma and Thailand (Taylor & Alexander, 2005).

Several studies based on sporocarp survey (Lee & Kim, 1987; Molina *et al.*, 1992; Newton & Haigh, 1998) have revealed the ecological specificity and host ranges of a variety of EM fungal species. As reported by Ishida *et al.* (2007), the absence of sporocarps does not necessarily indicate a lack of colonization and furthermore the approaches using sporocarps are problematic to understand the host specificity, when different host species are in close vicinity or occur in different field conditions. In contrast to the sporocarp approaches, the molecular methodologies can be applied to EM fungal species on individual host species in close vicinity within the same site, and to better understand EM fungal host specificity some authors (Ishida *et al.*, 2007) examined EM occurrence at multiple host pair taxonomic levels. They found a tendency similar to the some sporocarp studies, which suggested host specificity at higher levels of the host taxon (i.e. genus or family). This phenomenon is relatively common (Molina et al., 1992; Newton & Haigh, 1998; Massicotte et al., 1999). This may indicate that occurrence of EM fungal species is more common at the host family level than at the host species level, but this pattern may also be confounded by differences in statistical power among the host taxa compared (Ishida et al., 2007). Despite of the low level of colonization observed on host taxa, host preference and specificity were found for a considerable portion of EM species, suggesting that the presence of a variety of host taxa contributes to such EM fungal hyperdiversity in mixed conifer-broadleaf forests. This result supports the hypothesis that host diversity contributes to EM fungal diversity (Nantel & Neumann, 1992; Kernaghan et al., 2003). In addition, EM communities differed significantly among codominant tree species, indicating EM fungal spatial heterogeneity (Ishida et al., 2007). The EM fungal colonization on seedlings is related to the surrounding EM communities (Cline et al., 2005); moreover EM fungal communities can have different effects on different host species (Jonsson et al., 2001). Therefore the heterogeneity of the EM fungi may contribute to the establishment of various host species and the ectomycorrhizal fungal hyperdiversity in mixed conifer-broadleaf forests, may be maintained by this host diversity. The coexistence of various host species may in turn be supported by diverse and spatially heterogeneous EM communities (Ishida et al., 2007). As van der Heijden et al. (1998) have already demonstrated a positive influence of endomycorrhizal (VAM) diversity on plant diversity, the possibility then exists that EM diversity in mixed-wood forests may be maintained by a positive feedback between plant and fungal communities.

Plant species without mycorrhiza are mainly restricted to taxonomically defined groups of plants, such as the *Cyperaceae*, *Caryophyllaceae* and *Brassicaceae* families, or confined to aquatic or saline habitats (Harley & Harley, 1987).

3. The ectomycorrhizal mycobionts

Saprotrophic and mycorrhizal fungi are not separate groups from an evolutionary perspective, because the ability of fungi to form symbiotic associations with plants is a life strategy that has appeared from ancestral saprotrophic strategies during the evolutionary history (Hibbett et al., 2000). The most part, ca. 95%, of EM fungal species are homobasidiomycetes; the remaining species being ascomycetes (4.8%) and a few zygomycetes within the genus *Endogone* (Molina et al., 1992). Despite that a recent study by Weiss et al. (2004) demonstrated the understimated importance of heterobasidiomycetes with the *Sebacinaceae* as mycorrhizal formers, which have been strongly implicated as the mycobionts in EM (e.g. Urban et al., 2003), orchid (e.g. Taylor et al., 2003) and ericoid mycorrhizas (Allen et al., 2003). The recent description of mycorrhizal associations in jungermannioid liverworts also seems to involve members of this family (Kottke *et al.*, 2003).

Larsson *et al.* (2004) recognised how widespread the homobasidiomycetes are as EM formers with seven clades containing EM taxa. Anyway their fruitbodies formed by EM fungi are consequently very different and include thin, crust-like (resupinate), coral-like (clavarioid), cantharelloid, and agaricoid as well as boletoid structures. The majority of EM species are euagarics, and many of the most frequent and familiar sporocarps (e.g. *Amanita* spp.) that appear in forests in the autumn are formed by EM taxa (Taylor & Alexander, 2005).

If a genus is mycorrhizal, it does not mean that the fungi can form only ectomycorrhiza: as a single fungal species can form ecto- and arbutoid mycorrhizas (Smith & Read, 1997) on different host species (Horton *et al.*, 1999).

At one time the genus *Paxillus* was considered to be an exception as it was thought to contain both EM formers (*P. involutus* and *P. rubicundulus*) and saprotrophic species (e.g. *P. atromentarius* and *P. panuoides*) (Taylor & Alexander, 2005), but now the latter species belong to the saprotrophic genus *Tapinella*.

The knowledge of the ecology of ascomycete EM fungal species is very limited with the exception of some some *Tuber* spp. (Murat *et al.*, 2004). The importance of the *Helotiales* as EM mycobionts is suggested in recent work (Vrålstad *et al.*, 2000, 2002) up to now not valued. The identification of EM fungal species on short roots is a difficult task, but thanks to the molecular studies the accuracy of the classification has been greatly improved and the number of symbionts increases, so will the taxonomic range of the identified mycobionts (Taylor & Alexander, 2005).

For these reasons, up to now, the use of a combination of anatomical and molecular identification techniques is the most reliable method to study ectomycorrhizal community. In addition, a number of fungal groups considered to be saprotrophic in the past, have been found to be EM [e.g. tomentelloid fungi (Kõljalg *et al.*, 2000)]. In fact until recently, only a few genera of resupinate fungi with few species have been considered to be ectomycorrhizal, i.e. *Tylospora* (Eberhardt *et al.*, 1999), *Piloderma* (Larsen *et al.*, 1997), *Amphinema* (Fassi & de Vecchi, 1962), and *Byssocorticium* (Brand 1991).

The total number of EM fungi is very unclear and likely underestimated, but the most recent estimate (Molina *et al.*, 1992) suggested that there were about 5500 species. Thanks to the recent more intensive mycological explorations of tropical forests (e.g. Haug *et al.*, 2005; Buyck *et al.*, 1996), and of the hypogeous fungi associated with the eucalyptus vegetation in Australasia (Claridge 2002), many unknown EM species were discovered. In summary, an accurate estimation of the size of the global community of EM fungi may not be known for some time but it is likely to be about 7000 – 10.000 species (Taylor & Alexander, 2005).

The geographical distribution of the major mycorrhizal species in natural ecosystems has been suggested to follow altitudinal and latitudinal ecological gradients (Read 1991; Read & Perez-Moreno, 2003). This is probably explained by the change in factors limiting for plant growth, i.e. specific mycorrhizal associations are part of plant and fungal strategies to survive in various environments (Michelsen *et al.*, 1996; Cornelissen *et al.*, 2001). Furthermore, within each mycorrhizal anatomotype there might be a selection of specific fungal associates along ecological gradients (Taylor *et al.*, 2000; Lilleskov *et al.*, 2002; Read *et al.*, 2004). For instance the large number of EM fungi varies widely in capabilites to enzymatically attack organic polymers for capture of N and P (Leake & Read, 1997), and in their construction of the external mycelium (Agerer 2001). The significance of EM fungal functional diversity for ecosystem function, is still generally unexplored, particularly towards the Arctic and alpine regions (Clemmensen 2006).

4. The reasons of the investigations on the ectomycorrhizal communities
4.1 The ectomycorrhiza as bioindicator

Ectomycorrhizae, due to their key position at the plant-soil interface, are important to consider in the study of human disturbances like global change, the effects of pollution or forest management practices (Rillig *et al.*, 2002, Erland & Taylor, 2002).

From this point of view the ectomycorrhizae can be considered as a bioindicator, because an environmental indicator should reflect all the elements of the causal chain that links human activities to their ultimate environmental impacts and the societal responses to these impacts (Smeets & Weterings, 1999).

As reported by the work of Niemi & McDonald (2004), the ecological indicators have been applied in many ways in the context of both natural disturbances and anthropogenic stress. However, their primary role is to measure the response of the ecosystem to anthropogenic disturbances, but not necessarily to identify specific anthropogenic stress(es) causing impairment (US EPA, 2002a). Each ecological indicator referred as "state indicator", responds over different spatial and temporal scales; thus, the context of these scales must be explicitly stated for each ecological indicator. Understanding the response variability in ecological indicators is essential for their effective use (US EPA, 2002b).Without such an understanding, it is impossible to differentiate measurement errors from changing conditions, or an anthropogenic signal from background variation. In addition, they should be sensitive enough to react in a detectable way when a system is affected by anthropogenic stress, and they should also remain reasonably predictable in unperturbed ecosystems coupling with economic and social indicators. Legislatively, mandated use of ecological indicators occurs in many countries worldwide and is included in international accords. In Figure 2, the instruments, the planning steps at the basis of the ecological indicators application are summarized.

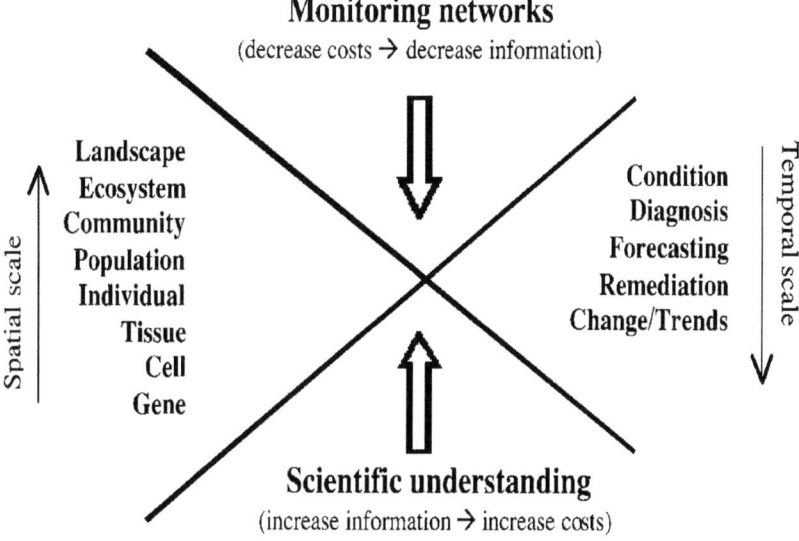

Fig. 2: Illustration of the suite of ecological indicators (*left*) for which a suite of assessment capabilities (*right*) are desired. Constraints on the development of ecological indicators at all levels for all assessment endpoints are due to a lack of scientific understanding and the predominance of policies requiring low cost monitoring. Goals in applications generally include a compromise between cost-effectiveness and the ability to defend the ecological indicator scientifically at the spatial and temporal scale appropriate to answer the desired management objectives: from Niemi & McDonald (2004).

Most ecological monitoring programs using ecological indicators are based on aggregated selected sites and on communities or population. But to recognize a particular trend in the ecosystem, an appropriate statistical design is necessary to understand the anthropogenic change against a background of natural variability. In fact, community and population respond to many other factors, some of which are not necessarily stress or stress-related. Furthermore, the researchers need to recognize which part of the ecological indicator spectrum is relevant to the objectives of their investigation. The applications of ecological indicators have focused at the species level like the studies on the EM communities, because the measurement assumes that a single species represents many species with similar ecological requirements, enforcing the definition of Landres *et al.* (1988). This is important to understand the definition of "focal species" used in the literature (Cox *et al.*, 1994; Lambeck 1997; Carroll *et al.*, 2001) but the concept has been expanded for use in conservation and management. The focal species represent those selected as a focus for a specific investigation (Niemi & McDonald, 2004) and have been used to identify potential indicator species, when there is a desire to describe ecological condition or measure the response to a disturbance. The failure of the measure was likely attributable to the narrow geographic ranges and restricted habitat distribution of rare species. Hence, information on rare species and those that are at risk was essential, yet gathering data on rare species is generally difficult, time-consuming, and expensive. In contrast to the indicator species approach, Manley *et al.* (2004) evaluated an innovative, multi-species monitoring for animals.

Historically, ecological indicators were primarily based on parameters associated with individual species (e.g., presence) or simple community metrics (e.g., species richness or diversity). However, many of these indicators did not fully represent the entire biological community of organisms present. Researchers have developed other indexes to provide more holistic approaches to ecological condition (Niemi *et al.*, 2004). These indexes range from simple diversity indexes, such as the Shannon and Wiener Index (Shannon & Weaver, 1949), to multimetric indexes (Simon, 2003). Multimetric ecological indicators are sets of mathematically aggregated or weighted indicators (US EPA ,2000, Kurtz *et al.*, 2001) that combine attributes of entire biotic communities into a useful measure of condition (US EPA, 2002b). Many other multimetric indexes have evolved over the past 20 years and in contrast to them, multivariate indexes (Reynoldson *et al.*, 1997) are statistical analyses of the biological community using a host of multivariate techniques, such as principal components analysis (O'Connor *et al.*, 2000), canonical correspondence analysis (Kingston *et al.*, 1992), and combinations of multivariate analyses (Dufrene & Legendre, 1997).

4.2 The methods applied for the studies

In a natural ecosystem the studies of mycorrhizal symbioses are carried out at multiple levels from the molecular level to the community complexity. The ectomycorrhizal fungal communities have traditionally been studied by surveying aboveground sporocarps identified using standard taxonomic approaches (Gardes & Bruns, 1996). The criticism of these approaches was clear, because these surveys do not always reflect the species composition of below ground EM fungal communities on root (Gardes & Bruns, 1996). The reasons were related to the lack of some EM fungi in aboveground surveys because they produce small or cryptic sporocarps, have no known sexual state or produce sporocarps infrequentely (Sakakibara *et al.*, 2002). Also, responses to environmental changes of EM communities on roots is probably delayed in comparison with responses in sporocarp communities, which further highlights the need to perform studies of EM fungal communities on roots (Wallenda & Kottke, 1998).

The EM communities are today mainly described by two methods associated with root tips: the molecular techniques and the morphological classification. The first method is based on the observation that the internal transcribed spacer (ITS) region of the nuclear rDNA exhibits a high level of variability among EM fungal species and minimal variation within species (Gardes & Bruns, 1996). After PCR amplification of the ITS region with fungi-specific primes, restriction fragment length polymorphism (RFLP) generated by enzyme digests of the ITS region can be used to separate several fungal species (Gardes & Bruns, 1996; Sakakibara *et al.*, 2002). RFLP patterns or ITS sequences are compared to databases with data on known fungal species, e.g. GenBank. Hence, these DNA-based techniques make it possible to identify the fungus with reasonable certainty and facilitate evaluation of both intra-and interspecific (Dahlberg 2001). However, the success rate of DNA extraction and amplification can vary among fungal species, and the use of DNA techniques on randomly sampled root tips, with no prior morphological categorization, cannot be used to generate quantitative descriptions of EM fungal communities (Sakakibara *et al.*, 2002).

Sometimes the morphological classification, "morphotyping" of the root tips with EM formations, needs different accuracy and throughput of a lot of sample material. In this thesis the morphotyping was made accurately, and not only confined to the morphological investigations. The morphological investigation was united to a more detailed anatomical observation and it was only the first step of this work, to avoid potential discrepancies.

Some fungi that can be distinguished from each other by DNA analyses have an indistinguishable morphology (Sakakibara *et al.*, 2002), however sometimes the sequences in the Genbanks are lacking or conflicting with the anatomical classification, as for the *Ramaria* genus or the Thelephorales members as reported in this thesis. For these reasons the morphotypes in this thesis were differentiated into anatomotypes [= species of ectomycorrhizae, according to Agerer *et al.*, (2002)]. It is therefore necessary to find a compromise between the accuracy of the method used to quantify fungal communities and the number of replicates needed to perform the statistical analyses, in order to obtain descriptions of heterogeneous EM fungal communities in plant roots in replicated multi-factorial field studies (Clemmensen 2006).

Thanks to the work of Tedersoo *et al.* (2007), new frontiers are offered for the future of the molecular analyses and consequentely for the phylogenetic researches. To overcome the problem of the contamination by misidentified and chimeric sequences accounts in the public sequence databases, the Nordic-Baltic initiative created the UNITE database (http://unite.ut.ee/) that includes well-annotated and vouched specimens identified by a taxonomist (Kõljalg *et al.*, 2005).

The data analysis is another important step to understand the quantitative response of the EM communities to the ecological factors. The diversity measures including richness and evenness are usually compared using conventional statistics.

As reported by Tedersoo et al. (2007) compositional data is best analysed using various ordination methods, the choice depending on hypotheses and software, but the ordination results usually provide some implications whether the community composition as a whole changes and which factors account for most of variation. Moreover, the ordination itself proves nothing because most methods lack relevant statistical testing and alternative ordination methods or distance algorithms can produce contrasting results. Furthermore, similar problems exist also during the result interpretations, because the species' position relative to the axes and factors provide a sound basis for developing new hypotheses that could be subsequently experimentally tested. The Detrended Correspondence Analysis (DCA) and Canonical Correspondence Analysis (CCA) are among the most consistent and useful ordination methods for indirect and direct gradient analysis. The PC-Ord (McCune & Mefford, 1999) or CANOCO software (ter Braak & Šmilauer, 2002) are the most sophisticated and demanding statistical tools up to now.

A great contribution to the ecological studies on EM fungal community was the work of Taylor (2002), highlighting the importance of the sampling to complete the diversity assessment, discussing the physical sampling strategy employed and the life cycle traits of the EM fungi being examined. As reported in these researches the structure of most EM communities is based on the presence of few common species and a large number of rare species. The laws at the basis of the theoretical detection limits showed the need to understand the sampling effort involved in assessing species richness.

The most recent techniques applied to the EM species are the isotopic tracer studies, following the fate of added stable or radioactive isotopic tracers (e.g. ^{15}N, ^{13}C, ^{14}C, ^{32}P), through different ecosytem pools can be used to assess fluxes of specific substances in an ecosystem. Isotopic labels have been widely used in laboratory studies to demonstrate uptake of different N-forms and to determine N- uptake kinetics in plants (Kielland 1994; Taylor et al., 2004), ectomycorrhizal fungi (e.g. Lipson et al., 1999). Isotopic labels have also proved valuabel tools to determine the rates of specific soil processes in field settings, e.g. nitrification (Stark & Hart, 1997). Applying $^{13}CO_2$ $^{14}CO_2$ tracers, among many other results, have given some insight into flux rates of C from plants to mycorrhizal fungi in laboratory studies (Jones et al., 1991; Heinonsalo et al., 2004). Also $^{13}CO_2$ or $^{14}CO_2$ tracers demonstrated reciprocal interspecific transfer of C among plants through common mycorrhizal mycelial networks (Simard et al., 1997). When applying isotopic tracers to natural ecosystems the most challenging task is to harvest and separate the pools that are analysed for the label (Clemmensen 2006). Because of naturally occurring $^{15}N:^{14}N$ and $^{13}C:^{12}C$ isotopic ratios, i.e. the ^{15}N and ^{13}C natural abundance, in various ecosystem pools can be used to reconstruct diet and trophic relationships as well as energy and mass flows within ecosystems (Post 2002). This is because most biochemical processes fractionate against the heavier isotopes so that the product has a lower isotopic ratio than the source of a process, e.g. N excreted from an organism is isotopically lighter than N kept (Clemmensen 2006). Thus, natural abundance of ^{15}N and ^{13}C of a pool or an organism reflects isotopic signatures of the inputs and outputs as well as the input-output balance. Notwithstanding the difficult sampling of different pools in an ecosystem, one advantage of the studies of natural isotopic abundance in comparison with isotopic labelling studies is that the method is not-manipulative.

Recent studies of ^{15}N and ^{13}C natural abundance were carried out in sporocarp communities in forest ecosystems and they have identified a difference between EM and saprotrophic (SAP) fungi linked to their trophic status (Högberg et al., 1999; Hobbie et al., 2001; Henn & Chapela, 2001; Taylor et al., 2003; Trudell et al., 2004). EM fungal species are relatively enriched in ^{15}N and ^{13}C compared to saprobes and autotrophs (Gebauer & Dietrich, 1993; Taylor et al., 1997) due to different N and C sources (Gleixner et al., 1993; Högberg et al., 1999), as reported in these previous studies on sporocarps. SAP fungi show the attitude to degrade dead organic material and use complex C sources like cellulose and lignin, whereas EM fungi receive more simple carbohydrates directly from the plant(s) with which they associate (Cooke & Whipps, 1993; Smith & Read, 1997). On the other hand the SAP and EM fungi have access to the same N sources in the soil, and the ^{15}N enrichment of EM relative to SAP fungal tissues is hypothesized mainly to be related to fractionations of N isotopes within the EM mycelium and preferential transfer of ^{15}N-depleted compounds to the plant (Högberg et al., 1996).

It needs to be emphasized that stable isotope concentrations are taxonomically biased within "functional guilds" both in plants (Delwiche et al., 1978) and fungi (Taylor et al., 2003). This basis should be considered when choosing reference taxa and comparing across temporal and spatial scales (Taylor et al., 2003). Ignoring these facts may lead to incorrect conclusions, especially when assigning trophic status of fungi (Tedersoo et al., 2007).

Stable isotopes were used in this thesis, to assign the trophic status of a species that belongs to the *Hygrophorus* genus (chapter 3). The use of stable isotope techniques in plant ecological research has grown steadily during the past two decades and this trend will continue as investigators realize that it can serve to understand the plant-environment interactions (Dawson et al., 2002).

4.3 The ectomycorrhizal communities in the soil

It is well-known that EM fungal communities frequently have a high species richness, in some cases exceeding 100 taxa in relatively small plots of land (Izzo *et al.*, 2004), and many species of EM fungi coexist in a mosaic fashion in a small volume of soil (Zhou & Hogetsu, 2002). Most comprise few, frequently occurring species and many more rare species (Taylor, 2002; Bueè *et al.*, 2005; Koide *et al.*, 2005). Species may spatially partition the forest floor (Dickie *et al.*, 2002; Genney *et al.*, 2006) and interact with each other both positively and negatively (Agerer *et al.*, 2002; Koide *et al.*, 2005).

The relationship between the frequency of soil hyphae, presence and numbers of fruiting structures and colonized roots change substantially among species (Gardes & Bruns, 1996; Gehring *et al.*, 1998; Koide *et al.*, 2005).

Significant variation among species, also relieved in their enzyme activities (Bueé *et al.*, 2005; Courty *et al.*, 2005; 2006), may explain in part why species vary in their capacity to absorb and transport N or P to their hosts, or in their demand for host C. Such enzyme assays may be especially relevant for species of ectomycorrhizal fungi that possess the contact type of hyphal exploration strategy (Agerer 2001). For other species, the hyphae growing into the soil may be at least as important to nutrient capture as colonized roots (Koide *et al.*, 2007).

Although the highest fine root density in boreal forest soils is found in the organic and upper mineral soil horizons (Persson 1980; Sylvia & Jarstfer, 1997; Makkonen & Helmissari, 1998), tree roots can be found at greater depths (Jackson *et al.*, 1996).

At all soil depths, fine roots are colonized by ectomycorrhizal fungi (Egli 1981). Most of the ectomycorrhizal fungal community studies restricted sampling to the upper, organic part of the soil profile (Horton & Bruns, 2001), ignoring the ectomycorrhizal root tips in the deeper mineral soil layers.

Chemical and mineralogical properties of soils change with depth, creating a number of different habitats for microrganisms, and the ectomycorrhizal fungal community is likely to change throughout the soil profile (Rosling *et al.*, 2003).

Studies on the distribution of ectomycorrhizal taxa in soil suggested that there may be large differences in species composition between the organic layer and the mineral soil (Egli 1981; Goodman & Trofymow, 1998; Fransson *et al.*, 2000; Danielsson & Visser, 1989; Heinonsalo *et al.*, 2001). Dickie *et al.* (2002 using T-RFLP analysis of DNA extracted from soil mycelium, found differences in ectomycorrhizal species composition between different components of the forest floor (L, F and H layers) and the B horizon of the mineral soil in a North American *Pinus resinosa* stand, while Zhou & Hogetsu (2002) used T-RFLP to map the three-dimensional distribution of ectomycorrhizal root tips in a Japanese *Larix kaempferi* stand, but found no clear vertical distribution patterns. Abuzinadah & Read (1986) suggested that the fungi found in the organic layers were adapted to using organic nutrients, while those in the mineral soil were more dependent upon mineral N. Further, Conn & Dighton (2000) and Dighton *et al.* (2000) demonstrated the importance of the litter chemistry in determining the species composition of EM fungi colonizing litter patches, as confirmed also by Toljander *et al.* (2006).

In conclusion, the species composition of EM fungal communities can be strongly influenced by various soil properties, including parent material (Gehring *et al.*, 1998; Scattolin *et al.*, 2007), soil stratification (Malajczuk & Hingston, 1981; Dickie *et al.*, 2002; Landeweert *et al.*, 2003; Rosling *et al.*, 2003), organic matter content (Harvey *et al.*, 1987), litter quality (Goodman & Trofymow, 1998; Conn & Dighton, 2000), moisture content (O'Dell *et al.*, 1999) and fertility (Sagara 1995; Lilleskov *et al.*, 2001). Variation in any of these soil variables has the potential to contribute to ectomycorrhizal fungal species diversity. The result of the abiotic preference of a species, which delineate its fundamental niche, can be at the basis of the kind of partitioning of the environment, as well as evident interactions among the same species could limit a species to its realized niche. These interactions among EM fungal species probably occur frequently (Koide *et al.*, 2005).

The findings of several studies suggested that there is frequent opportunity for at least some EM fungi within a community to interact with others as their hyphae attempt to colonize either newly produced roots in order to acquire carbon, or volumes of forest floor in order to capture water and nutrients (Gryta *et al.*, 1997; Fiore-Donno & Martin, 2001; Guidot *et al.*, 2001, 2003; Zhou & Hogetsu, 2002). Furthermore, many researchers have noted multiple species of EM fungi colonizing a single root (Mamoun & Olivier, 1993a, b; Wu *et al.*, 1999), and association between specific pairs of fungi on colonized roots, Olsson *et al.*, (2000). Unfortunately, there is limited knowledge as to whether there are special ecological microniches in the soil for morphologically different ectomycorrhizae (Agerer *et al.*, 2002).

Interactions among EM fungal species may be either positive (co-occurence, specific associations) or negative (competitive exclusion) in nature (Agerer *et al.*, 2002; Koide *et al.*, 2005). For example, the persistence may be related to the ability of a species to exclude others from colonizing roots (Fleming 1985; Mamoum & Olivier, 1993a, b; Olivier & Mamoun, 1994) and in some cases, this could involve the production of chemical inhibitory substance as for the species *Cenococcum geophilum* Fr. (Koide *et al.*, 2005). However, systematic investigation of such not random distributions at the whole-community level has not been made (Koide *et al.*, 2005).

Izzo *et al.* (2005) sampled at intervals ranging from 5 cm to 200 cm at two depths over 3 years.

With this combined approach the authors demonstrated that the ectomycorrhizal community turns over frequently at smaller scales, but much less so than at larger scales. These results indicated that the pool of available ectomycorrhizal species within an ecosystem may remain more or less constant, whereas the exact location of individual species may shift over time.

The assessments of EM communities are important because a growing body of research suggests that mycorrhizal species vary in their influence on a number of ecological processes (Treseder 2005).

Izzo et al. (2005), found that inter-annual variation in climate and other factors had little impact on the composition of the ectomycorrhizal community within the forest as a whole even over 3 years. By contrast, short-term changes pointed to the growth or death of particular individuals, rather than the loss or immigration of species. This distinction is important because it implies that the short-term changes may be easily reversible, depending on future conditions (Treseder 2005). In disturbed ecosystems the results attested a different situation, in which changes in ectomycorrhizal composition have been recorded within a few years after exposure to fire, elevated CO_2, or nitrogen additions (Grogan et al., 2000; Treseder & Allen, 2000). These different approaches are discussed to understand if these shifts in community composition are due to alterations in an otherwise stable "background" pool within the ecosystem, as might be implied by Izzo et al. (2005), or if they are a result of easily reversible changes in population dynamics (Treseder 2005).

4.4 The ectomycorrhizal responses to environmental stress

The factors that influence community development and maintain the high EM fungal diversity present in boreal ecosystems are poorly understood. Studies which have examined determinants of EM fungal diversity under natural undisturbed systems, don't abound in contrast to several studies examining diversity in relation to changes in abiotic factors due to pollution and/or forest management practise (Erland & Taylor, 2002).

The typical structure of a EM community consists of a few common species, colonising 50-70% of the available fine roots, and a large number of rare species (Buée et al., 2005; Erland & Taylor, 2002; Koide et al., 2005; Taylor 2002). The community diversity is usually considered to have two components: the number of the species or species richness and the relative abundance of species or community evenness (Magurann 1988). The high species richness, often reported in EM fungal community investigations, is due to several mechanisms that may contribute to including spatial and temporal partitioning, as a result of the edaphic variation or the interactions among the species (Koide et al., 2007).

The most frequent response of the EM community to a perturbation due to anthropogenic factors is a shift in the community structure such as the dominance increases and species richness declines as reported by different studies (reviewed by Erland & Taylor, 1999; De Roman *et al.*, 2005; Mosca 2007), on Norway spruce (Kraigher 1999), oaks (Kovacs *et al.*, 2000). Pollution and other anthropogenic stresses have been found to diminish biodiversity indices of EM also in spruce stand in Slovenia by Kraigher *et al.* (2006); however in European beech this trend was not detected (Kraigher *et al.*, 2006).

New species may appear more important as colonisers after a disturbance (Kraigher *et al.*, 2006). However Erland & Taylor (2002) suggested that until the spatial distribution is not really understood and the sampling strategies are not well developed to deal with the non-random distribution of the EM fungi in soil, the results could suggest changes in the community diversity, which should be interpreted with caution. The sampling strategies must be able to accommodate the changes of root tip density following a perturbation, because the number of root tips in a sample can significantly affect the number of EM fungal species found.

Tab. 1: **Summary of known effects of management and pollution upon the EM community (from Erland and Taylor 2002 with modification).**

Factor	Mycorrhizal tip numbers	Colonisation (%)	Sporocarp production	EM community belowground	Extramatrical mycelium	General comments
Elevated CO_2	Increase in fine root production often recorded (Rey and Jarvis 1997; Runion et al. 1997)	No effect recorded	No data	Changes in species composition in pot cultures (Godbold et al. 1997; Rey & Jarvis, 1997)	Increased production (Godbold et al. 1997; Rouhier & Read 1998, 1999)	Insufficient data, particularly field data
Ozone	Possible decrease (Edwards & Kelly, 1992)	No effect (Roth & Fahey, 1998)	No data	Change in community structure (Edwards and Kelly 1992; Qiu et al. 1993)	Insufficient data	Insufficient data

Heavy metals	Effects dependent upon metal spp. and conc. (Hartley et al. 1999)	Effects dependent upon metal spp. and conc. (Hartley et al. 1999)	Decrease. Species richness negatively affected (Rühling & Söderström, 1990)	Increase in tolerant species (Hartley et al. 1999)	Insufficient data	Larger inter-and intraspecific differences. Complex interactions between plant/fungus/metal (Leyval et al. 1997). Percent colonisation may decrease, especially if host is more tolerant than mycobionts
	Decreases have been reported (Kraigher et al. 1996; Erland et al. 1999)	No change (Wallenda & Kottke 1998; Taylor et al. 2000)	Initial change in community structure. Reduction in sporocarp production (Baar and ter Braak 1996) but increased production by tolerant spp. may mask decrease in sensitive spp. (Wallenda & Kottke, 1998)	Decrease in diversity (Kraigher 1996; Lilleskov and Fahey 1996; Taylor et al. 2000) decrease in protein spp. (Taylor et al. 2000)	Insufficient data	Decrease in diversity, both in terms of spp. Richness and evenness. More severe effect above ground. "Specialist species" more adversely affected. (Wallenda & Kottke, 1998). Deficiency of other nutrients may lead to higher numbers of EM
N-fertilisation	Short-term decrease after large single N additions (Meyer 1962; Ahlström et al. 1988)	Short-term decrease after large single N addition (Wallenda & Kottke, 1998)	Differential response- some spp. increase, e. g. *Lactarius rufus*, most spp. decline	Insufficient data, particularly with regard to long term effects.		Most studies record decreasing diversity after large single N additions

			(Wallenda & Kottke, 1998)	Changes in community structure recorded (Arnebrant and Söderström 1992; Kåren and Nylund 1996)		
Acidification	Decrease in fine root numbers (Dighton & Skeffington, 1987)	No change (Dighton & Skeffington, 1987)	Decline in diversity (Arnolds 1991; Dighton and Skeffington 1987; Agerer et al. 1998). Increased production by acidophilous spp. (Agerer et al. 1998)	Changes in species composition (Roth & Fahey, 1998; Qiu et al. 1993). Decreases in spp. with abundant extramatrical mycelium	Decreased production (Dighton & Skeffington, 1987)	Increased disturbance due to greater earthworm activity could reduce Extramatical mycelium
Liming	Often large increase in root tips (Erland & Söderström, 1991; Persson & Ahlström, 1994; Jonsson et al. 1999; Bakker et al. 2000)	No change recorded but few data available on immediate effects of liming	Differential response by spp. (Agerer et al. 1998)	Considerable changes in spp. Composition often recorded after liming. (Lehto 1984, 1994; Erland & Söderström, 1991; Andersson & Söderström, 1995; Jonsson et al.1999)	Increase in types with abundant mycelia (Bakker et al. 2000)	There is a great need for more studies into the effects of liming

Wood ash	Insufficient data	No change (Mamood, 2000)	Insufficient data	Some evidence of differential spp. Response (Mahmood, 2000)	Insufficient data. EM nycelia have been reported to colonise ash granules (Mahmood, 2000)	Very few studies available
Vitality fertilisation	Insufficient data	No change Kåren & Mylund, 1996)	Insufficient data	Insufficient data	No data	Very few studies

In Germany recent studies were performed on mature beech and spruce to discuss a possible ecological role for the abundant types of ectomycorrhiza and their putative application in ozone impact bioindication (Grebenc & Kraigher, 2007). The total number of mycorrhizal fine roots was higher at the fumigated plot as compared to the control site. Some species as *Cenococcum geophilum* Fr., *Russula densifolia* Romagn., *Russula fellea* Fr. (Fr.), *Russula illota* Romagn., *Tuber puberulum* (Berk.) Broome were more abundant under ozone-fumigated trees, and other species like *Lactarius acris* (Bolton) Gray, *Fagirhiza fusca* (Brand 1991) and *Fagirhiza setifera* (Brand 1991) were present only in fumigated plots.

Coleman *et al.* (1992) described the soil as the "chief organizing centre for ecosystem function". The role of soil biota and processes as modifiers of the ecosystem or plant responses to global change is becoming increasingly recognized. One of the main functions of mycorrhizal fungi and fungi in general at the ecosystem level is their contribution to the formation and maintenance of soil structure (Tisdall & Oades, 1982). The global change factors can influence other soil biota, physical features (soil structure) and can have potentially large indirect effects on the EM community composition of mycorrhizal fungi and mycorrhizal functioning (Rillig *et al.*, 2002). Consequently each global change effect on the extraradical mycelium of mycorrhizal fungi can secondarily impact soil structure (Young *et al.*, 1998; Rillig *et al.*, 1999).

The significance of shifts in EM fungal diversity at the ecosystem level remains unclear due to a lack of knowledge of the functional capabilities of most EM fungal taxa under field conditions. Up to now it is known, however, that considerable interspecies variation exists with regard to a number of physiological attributes for instance the nutritional host status can be affected by the changes in dominance (Erland & Taylor, 2002).

Due global change effects, the real interpretation of the shift is becoming increasingly difficult, which in turn make it more difficult to discern the result of each component, notwithstanding the many multidisciplinary researches. However, the climate change on Mediterranean forests is evident with a current biome shift, which has been documented recently (Penuelas & Boada, 2003), in which *Calluna vulgaris* (L.) Hull (Heather) and *Fagus sylvatica* L. are being replaced by *Quercus ilex* (L.) in higher elevations as it extends beyond the former upper limit of its range with onset of the milder weather conditions.

In this context, recent studies on global change, considering the predicted increase in drought frequency and intensity, reported that the short-term consequences of drought on biodiversity depend on species abilities to resist, and to recover after the drought, and on competitive interactions between species. Although the abundance of many species generally decreases during droughts, some taxa may increase in number during droughts or shortly thereafter (Archaux & Wolters, 2006) as reported for the EM communities responses to the general human pressure.

The work on different beech ecotypes of Shi *et al.* (2002) confirmed in drought causes a shift in plant/fungus communities, showing that decreased soil water availability did not significantly change either the degree of fungal colonisation of the roots, nor the number of ectomycorrhizal types per root system. Droughts did, however, have an influence on the composition of the ectomycorrhizal community. Different mycorrhizal types responded to droughts differently in terms of patterns of occurrence/abundance. Droughts increased the abundance of mycorrhiza formed between beech and *Xerocomus chrysenteron* (Bull.) Quél. Sustained partitioning of carbon towards the mycorrhizal fungi under drought was reflected by an increase of nitrogen storage in the fungal vacuoles (Shi *et al.*, 2002).

The temperature can have also direct effects on mycorrhizal fungi (Staddon *et al.*, 2002) as all organisms have an optimum for temperature conditions. The enzymatic activity depends on the temperature, since the latter can directly affect mycorrhizal fungi, and also due to its impacts on host plants. Temperature effects can also be indirect, via effects on other environmental factors (e.g. soil moisture).

As for elevated CO_2, it may be found that in natural ecosystems the effects of temperature on mycorrhizae will mainly be, due to temperature-induced changes to plant communities.

Furthermore the work of Izzo *et al.* (2006) is an important improvement to understand the EM responses to fire. They tested the behaviour of the resistant propagule community on heat treatments in artificial conditions. One species, *Rhizopogon olivaceotinctus,* significantly increased in frequency, and two species (*Cenococcum geophilum* and *Wilcoxina* sp.) significantly decreased in frequency after a 75° C treatment. The increase of *Rhizopogon olivaceotinctus* A.H. Sm., coupled with other features of its behaviour, suggests that substantial heat disturbances may benefit this species in competing for roots. But in natural ecosystems many soil properties (e.g. nutrient availability, pH, and hydrophobicity) are present in an altered form, resulting from heat and the drying stress of fire (Agee 1993). Moreover, these effects may vary across space and soil depth.

To partition the effects of these different factors, Grogan *et al.* (2000) examined the effect of removing post-fire ash, an important nutrient source, on EM community composition on field seedlings, but due to high species richness and spatial variability they found no clear effects. Baar *et al.* (1999) found that propagules of some EM species responded positively to soil drying in greenhouse experiments, and that these were the same species that colonized seedlings in nature following fire.

The importance of soil changes is also important to understand the sylvicultural impacts on EM communities. For instance, as reported by Baar & de Vries (1995), the manipulation of litter and humus layers strongly affects the ectomycorrhizal colonization capacity. Termorshuizen (1991) showed that the occurrence of ectomycorrhizal fruitbodies, in Scots pine forests of different ages, and the seedlings mycorrhization, is not linked to the aging of the trees, but to the aging of the forest soil, which is likely to be the main factor determining ectomycorrhizal infection. Based on the field experiment, it is also concluded, in the work of Heinonsalo (2004), that the shift in ectomycorrhizal community structure, observed in the seedling roots after clear-cut logging, is not due to the lack of inoculum in the clear-cut soil, but to changes in the soil environment.

Normally, the spatial richness decreases in EM diversity after clear-cuts (Cline *et al.*, 2005) or it can increase after a thinning (Bueè *et al.*, 2005; Mosca *et al.*, 2007). Harvesting significantly also decreased the thickness of the humus layer, as well as the numbers of ectomycorrhizal root tips, both per metre root length and per unit humus volume (Mahmood *et al.*, 1999).

The results on a particular type of repeated silviculture action are lacking up to now.

In chapter 4 of this thesis the results of studies on the EM community structures in beech coppices are documented, to understand the possible resilience or an adaptative diversity of the EM species as reported in previous investigations (Mosca 2007; Scattolin 2007).

The number of EM morphotypes increased with stand age along the chronosequence also in the studies of Gebhardt *et al.* (2007), performed on EM communities of red oak (*Quercus rubra* L.) of different age in the Lusatian lignite mining district, East Germany. However, the number of morphotypes was lower in stands with disturbed soil than with undisturbed soil and the age has probably influenced the colonization rate of red oak, because it was slower only in the youngest chronosequence stand.

The notion that the age of a stand of trees could influence the community structure of EM was first postulated in the early 1980s (Mason *et al.*, 1982; 1983). The concept generated considerable interest because Mason and his colleagues suggested that some species of EM fungi were found only when trees were in their pioneer phase, so-called early stage fungi and others were specific to climax vegetation, so-called late-stage fungi. This theory was later refined by Danielson (1984) to include a third category known as multi-stage fungi.

Numerous studies have utilized forests with a gradient of stand ages to test Mason's hypothesis with varying degrees of agreement. As a general point, it should be noted that several of these studies are compromised by the lack of true replication. This essential requirement for meaningful statistical analysis is not always easy to achieve in chronosequences. Moreover, many other factors are frequently correlated with stand age and careful experimental design, field observation, and statistical analyses are required to try and separate the various factors tested. The critics of the EM succession hypothesis have argued that the hypothesis is only likely to hold true for pioneer plant species because the original study utilized a stand of birch (*Betula pendula* Roth) that had recently colonized agricultural soil (Johnson *et al.*, 2005). Visser (1995) studied a chronosequence of a Jack pine (*Pinus banksiana* Lamb) that had regenerated naturally after wildfire disturbance. The data showed that the number of EM morphotypes increased progressively in the first 65 years before increasing at a much-reduced rate until 122 years and the EM community included early-stage species such as *Coltricia perennis* (L.) Murill, multi-stage species such as *Suillus brevipes* (Peek) Kuntze, and late-stage species such as *Cortinarius* spp.

Furthermore, the view of the EM community did not show the predicted decline in species richness following canopy closure (Last *et al.*, 1987). A similar trend was seen in stands of *Pinus kesiya* Royle ex Gordon during the initial (2–17 year) growth phase (Rao *et al.*, 1997). Here, species richness of EM fungi was directly proportional to the age of the stand. All these cited studies however are subject either to the vagaries of the relationship between sporocarp presence and mycorrhiza presence or from the uncertainties of EM morphotype identification (Johnson *et al.*, 2005).

Lee & Alexander (1996) obtained similar data for the EM fungi in tropical rain forests and demonstrated that EM community on the roots of dipterocarp seedlings changed in the 7 months following germination, showed that fungi entered the mycorrhizal community as time progressed, some fungi were lost or declined in relative abundance, and regarded this as a clear evidence of succession (Johnson *et al.*, 2005). As far as known, there have yet to be any studies using DNA-based methods to investigate the successional processes in EM communities, despite the recognition 10 years ago that this would be a useful line of inquiry (Egger 1995). Species richness in the surface organic (L, F, and H) horizons was the least in the youngest stand (13 years) and was the greatest in the 59 and 116 year-old-stands. By contrast, species richness in the mineral horizon (in this case a uniform sand horizon several metres deep) did not differ between stand ages. This data suggests a rather idiosyncratic response of root-associated basidiomycete fungal communities to host plant age. It is clear that tree age can have impacts on EM fungal communities, but that these may be more or less apparent in particular forest types, notably young plantations versus old growth (Johnson et al., 2005). There seem to be two processes occurring: changes in mycorrhizal communities on individuals with time at that individual inocula are available, and also changes at the stand level associated with a range of edaphic factors. Several authors have alluded to the latter point (Johnson et al., 2005). Visser (1995) highlighted that differences in host carbon supply could have driven the changes seen in the EM fungal communities. This hypothesis arises from the notion that carbohydrate supply can affect EM colonization (Björkman 1949). The isotope tracer techniques required to determine if EM community composition is related to host carbon supply are readily available and have been highlighted already.

It is further deducible that the extramatrical mycelium is likely to be the component of the belowground EM community that is most sensitive and responsive to environmental change (Erland & Taylor, 2002). In the recent years new advances (Anderson & Cairney, 2007) in understanding soil-borne mycelia of EM fungi have arisen from combined use of molecular technologies and novel field experimentation. These approaches have the potential to provide unprecedented insights into the functioning of EM mycelia at the ecosystem level, particularly in the context of land-use changes and global climate change. EM fungal mycelia can comprise 80% of the total fungal biomass and 30% of the microbial biomass in some forest soil (Wallander et al. 2001, 2003; Högberg & Högberg, 2002), with carbon allocation to EM fungi estimated to be as much as 22% of net primary production (Hobbie 2006). EM fungi are thus an important component of forest carbon cycles, and the effects on elevated atmospheric CO_2 have received more attention in the last years (Anderson & Cairney, 2007). Elevated atmospheric CO_2 conditions showed increased percentage root colonization by EM fungi (Norby et al., 1987; Ineichen et al., 1995; Berntson et al., 1997; Godbold et al., 1997; Tingey et al., 1997; Rouhier & Read, 1998; Walker et al., 1998; Kasurinen et al., 2005), with only one exception (Rouhier & Read, 1999). Confirming this response some effects were also shown on the altered EM fungal root-tip community structure in experimental conditions (Norby et al., 1987; Ineichen et al., 1995) and in the field (Fransson et al., 2001; Kasurinen et al., 2005). A change in EM root-tip community composition in favour of morphotypes that appeared to produce emanating hyphae and/or rhizomorphs was noted under these conditions (Goldbold & Bernston, 1997; Goldbold et al., 1997). Because the attempt to quantify soil-borne mycelia is lacking, this results can't provide direct information on the mycelia response of EM fungi. Another criticism is the inability to generalize, because the results are available only on single field experiments and on the response of the EM mycelial community and, therefore, they provide no information on the behaviour of individual EM species to elevated atmospheric CO_2 in the field (Anderson & Cairney, 2007).

5. The beech root system and its mycorrhizal root structure

The root system of *Fagus sylvatica* L. has been described by Büsgen (1905) as intensive because of the relatively large number of fine roots per unit volume.
The structure and mode of growth of the root tips of adult trees have been described in details by Clowes (1949, 1950, 1951, 1954). During their collection and dissection for physiological investigation Harley (1948) observed that the root system of the beech colonizes the soil within the immediate area of the tree canopy and produces a large number of fine roots in the surface layer of the soils The accumulation of rootlets near the soil surface is especially evident under woodland conditions. There appears to be a differentiation of the ultimate laterals into "long" and "short" roots; the short roots are often termed the "feeding roots" and the main function of absorption is ascribed to them. Indeed, as their surfaces constitute the greater part of the surface area of the root system, this must be true. It should not be assumed, however, that the long roots have solely a pioneer and anchoring function, for their apical regions are capable at least of absorbing salts and water (Harley 1948) and are often equipped with root-hairs. In *Fagus* there is no distinction in quality between the two types of roots, so that if a comparison is made of long and short roots that are uninfected by fungi, the types grade into one another. Harley (1948) was also the first to investigate the primary root, considering them in order of increasing intensity of infection.
The first mycorrhizas of *Fagus sylvatica* L. were recorded, in fact, by Harley (1948) and Warren Wilson (1951), as being after the first foliage leaves were set. Boullard (1960, 1961) and Laiho & Mikola (1964) observed the same features in *Pinus sylvestris* L., *Pinus montana Salisb.* (= *Pinus cembra* L.), and *Picea abies* L. In all these cases there was a correlation of the initiation of infection with the probable onset of active photosynthesis, and this can help to choose the right sampling time.
The results of Boullard (1961) on the effect of light on the infection of *Cedrus atlantica* L., *Pinus pinaster* Ait., *Pinus sylvestris* L. and other tree species, showed that an increase of the light period, i.e. Increase of the duration of the daily photosynthetic period, from 6 hours to 16 hours or even longer, increased the development of the root systems and the number of short roots on the seedlings. It also resulted in an increase in the number and percentage of roots on the seedlings. It also resulted in an increase in the number of roots converted to mycorrhizae Essentially similar

results were obtained by Wenger (1955) using *Pinus* and by Harley & Waid (1955) using *Fagus*. But this correlation is not always valid.

Clowes (1951) underlined the rare presence in *Fagus* of a tannin barrier outside the endodermis, as reported for other plants (MacDougal & Dufrenoy, 1946). Where it does occur it may take the form of a ring of cells (the epidermis or in the cortex) with droplets of tannin in the cytoplasm, or of a circle formed of tannin-impregnated cells (Clowes 1951).

As reported by Clowes, although the most common state of infection of *Fagus* is that normally described for ectotrophic mycorrhizae some of thicker roots (either long or short roots) had a mantle of fungal pseudoparenchyma without a Hartig net. The fungus laid on the surface of the epidermis or the outer cap cells and in a few cases the hyphae penetrated the outer cell wall into some of the epidermal cells and there formed swollen vesicles. More rarely intracellular hyphae ramified throughout the cortex and even into the meristem. This type of mycorrhiza occurred in the studies of Clowses frequently, when the infection is limited to the apex.

Götsche (1972) also found an ectoendotrophic status of mycorrhizae in mixed stands of *Picea abies* and *Fagus sylvatica*, where the species showed intracellular hyphae and progressively destroyed the epidermal cells. The formation of this parasitism form was connected with suberin presence.

The description of the EM on *F. sylvatica* began with the research of Brand (1991), who classified 23 new ectomycorrhizal species.

6. Aim of the thesis

The main goal of this thesis, linked to "InHumusNat2000" (1587/2004) project, funded by the "Fondo per i progetti di ricerca della Provincia autonoma di Trento", in co-operation with the Centre for Alpine Ecology (TN, Italy), was to verify the possibile response to the coppicing on beech (*Fagus sylvatica* L.), applying a biological indicator such as the ectomycorrhizal community could be.

These researches were performed in beech coppices of different age, because the last cut was applied in different periods. They were selected among sites, which are very important for the European Community for habitat and species protection ("SIC" sites or "ZPS" sites belong to the "Net Nat2000", in application of the European Directives n. 42 1992 and n. 409 1978, respectively). This work would integrate the parameters generally used for the management in forests of particular importance, but suffering under a hard and constant human pressure.

Since only a few studies on the vertical distribution of EM (ectomycorrhizae) consortium in soils are available up to now, this thesis is an attempt to associate the species composition in the different soil horizons.

To determine the influence of environmental features on the EM species distribution, pH, exposure, humus forms and their chemical-physical properties were taken into account as the most representative and influencing factors in soil ecological dynamics.

To establish the diversity and the EM community structure in these natural habitats the study was carried out in different stands, to relate the coppicing effects on the species in the soil layers.

Moreover, this thesis would be a small contribution to the biodiversity of the beech forest ecosystems with four new descriptions of ectomycorrhizae.

6.1 Thesis structure

The thesis is composed by five chapters presenting at first the composition of the ectomycorrhizal communities in beech coppices in the North of Italy, in the Trentino-Südtirol Region, reporting three new species descriptions (chapter 2).

Chapter 3 reports the investigation results on an ectomycorrhiza species, which shows a parasitic attitude in the studied sites.

Chapters 4 consists of two different articles, to describe the EM community response to the coppicing and in particular in the soil organic layers and related to the different environmental features. In chapters 5-9 five descriptions of new EM species are reported.

Each chapter is based on a paper submitted to, or in evolution for, an international peer-reviewed journal, then followed by a general conclusion (chapter 10).

References

ABUZINADAH RA, READ DJ (1986). The role of proteins in the nitrogen nutrition of ectomycorrhizal plants. Utilization of peptides and proteins by ectomycorrhizal fungi. *New Phytologist* 103: 481-493.

AGEE J (1993). Fire Ecology of Pacific Northwest Forests. Island Press, Washington D.C.

AGERER R (1987-2002). *Colour Atlas of Ectomycorrhizae.* 1st-12th delivery. Einhorn Verlag, Schwäbisch Gmünd.

AGERER R (1999). Comparison of the ontogeny of hyphal and rhizoid strands of *Pisolithus tinctorius* and *Polytrichum juniperinum*. *Crypt Bot* 2/3:85–92.

AGERER R (2001). Exploration types of ectomycorrhizae. A proposal to classify ectomycorrhizal mycelial systems according to their patterns of differentiation and putative ecological importance. *Mycorrhiza* 11: 107-114.

AGERER R, DANIELSON RM, EGLI S, INGLEBY K, LUOMA D & TREU R (1996 – 2004). *Descriptions of Ectomycorrhizae*. Schwäbisch-Gmünd: Einhorn-Verlag.

AGERER R, GROTE R, RAIDL S (2002). The new method "micromapping", a means to study species-specific associations and exclusions of ectomycorrhizae. *Mycological Progress* 1(2): 155-166.

AGERER R, SCHLOTER M, HAHN C (2000). Fungal enzymatic activity in fruitbodies. *Nova Hedwigia* Kryptogamenkd 71: 315–336.

AGERER R, TAYLOR AFS, TREU R (1998). Effects of acid irrigation and liming on the production of fruit bodies by ecto-mycorrhizal fungi. *Plant Soil* 199: 179-190.

AHLSTRÖM K, PERSSON H, BÖRJESSON I (1988). Fertilization in a mature Scots pine (*Pinus sylvestris* L.) stand – effects on fine roots. *Plant Soil* 106: 179-190.

ANDERSON IC, CAIRNEY JWG (2007). Ectomycorrhizal fungi: exploring the mycelial frontier. Federation of European Microbioloical Societies Blackwell Publisching 31: 388-406.

ANDERSSON S, SÖDERSTRÖM B (1995). Effects of lime ($CaCO_3$) on ecto-mycorrhizal colonisation of Picea abies (L.) Karst. Seedlings plantes in a spruce forest. *Scandinavian Journal of Forest Research* 10: 149-154.

ARCHAUX F, WOLTERS V(2006). Impact of summer drought on forest biodiversity: what do we know? Annal of Forest Science 63: 645–652 INRA, EDP Sciences.

ARNEBRANT K, SÖDERSTRÖM (1992). Effects of different fertilizer treatment on ecto-mycorrhizal colonization potential in two Scots pine forests in Sweden. *Forest Ecolgy and Management* 53: 77-89.

ARNOLDS E (1991). Decline of ecto-mycorrhizal fungi in Europe. *Agricolture Ecosytem and Environment* 35: 209-244.

ASHFORD, AE, PETERSON CA, CARPENTER JL, CAIRNEY JWG, ALLAWAY, WG (1988). Structure and permeability of the fungal sheath in the *Pisonia mycorrhiza*. *Protoplasma* 147: 149-161.

BAAR J, de VRIES FW (1995). Effects of manipulation of litter and humus layers on ectomycorrhizal colonization potential in Scots pine stands of different age. *Mycorrhiza* 5: 267-272.

BAAR J, HORTON TR, KRETSER AM, BRUNS TD (1999). Mycorrhizal colonization of Pinus muricata from resistant propagules after a stand replacing wildfire. *New Phytologist* 143: 409-418.

BAAR J, ter BRAAK CJF (1996). Ecto-mycorrhizal sporocarp occurrence as affected by manipulation of litter and humus layer in Scots pine stands of different age. *Application Soil Ecology* 4: 61-73.

BAKKER MR, GARBAYE J, NYS C (2000). Effect of liming on the ecto-mycorrhizal status of oak. *Forest Ecology and Management* 126: 121-131.

BERNTSON GM, WAYNE PM, BAZZAZ FA (1997). Belowground architectural and mycorrhizal responses to elevetated CO_2 in *Betula alleghaniensis* populations. *Functional Ecology* 11: 684-695.

BJÖRKMAN E (1949). The ecological significance of the ectotrophic mycorrhizal association in forest trees. *Svensk Botanisk Tidskrift* 43: 223–262.

BOULLARD B (1960). La lumière les mycorrhizes. *Annal of Biology* 36: 231-248.

BOULLARD B (1961). Influence du photopèriodisme sur le mycorrhizationde jeunes conifères. *Bull. Soc. Linn. Normandie, Série* 10 (2): 30-46.

BRAND F (1991). Ektomykorrhizen an *Fagus sylvatica*. Charakterisierung und Identifizierung, ökologische Kennzeichnung und unsterile Kultivierung. *Libri Botanici*, vol. 2, IHW-Verlag, Eching, Germany.

BUÉE M, VAIRRELES D, GARBAYE J (2005). Year-round monitoring of diversity and potential metabolic activity of ectomycorrhyzal community in a beech (*Fagus silvatica* L.) forest subjected to two thinning regimes. *Mycorrhiza* 15: 235-245.

BÜSGEN M (1905). Studien über die Wurzelsysteme einiger dictotyler Holzpflanzen. Flora, Jena 95: 58.

BUYCK B, THOEN D, WATLING R (1996). Ectomycorrhizal fungi of the Guinea Congo region. *Proceedings of the Royal Society of Edinburgh* 104B: 313-333.

CAIRNEY JWG, BURKE RM (1996). Physiological hetereogeneity within fungal mycelia: an important concept for a functional understanding of the ectomycorrhizal symbiosis. *New Phytologist* 134: 685-695.

CARROLL C, NOSS RF, PAQUET PC (2001). Carnivores as focal species for conservation planning in the Rocky Mountain region. *Ecological Applications* 11:961–80.

CLARIDGE AW (2002). Ecological role of hypogeousectomycorrhizal fungi in Australia forests and woodlands. *Plant and Soil* 244: 291-305.

CLEMMENSEN KE (2006). Ectomycorrhiza and Artic Ecosystem Response to Environmental Change PhD Thesis. Institute of Biology Faculty of Science University of Copenhagen.

CLINE ET, AMMIRATI JF, EDMONDS FRL (2005). Does proximity to mature trees influence ectomycorrhizal fungus communities of Douglas-fir seedling? *New Phytologist* 166: 993-1009.

CLOWES FAL (1949). The morphology and anatomy of the roots associated with ectotrophic mycorrhiza. D. Phil., Thesis, Oxford University (typescript).

CLOWES FAL (1950). Root apical meristems of *Fagus sylvatica. New Phytologist* 49: 249-268.

CLOWES FAL (1951). The structure of mycorrhizal roots of *Fagus sylvatica. New Phytologist* 50: 1-16.

CLOWES FAL (1954). The root-cap of ectotrophic mycorrhizas. *New Phytologist* 53: 525-529.

COLEMAN DC, ODUM EP, CROSSLEY DA (1992). Soil biology, soil ecology and global change. *Biology and Fertility of Soils* 14: 104-111.

CONN C, DIGHTON J (2000). Litter quality influences on decomposition, ectomycorrhizal community structure and mycorrhizal root surface acid phosphatase activity. *Soil Biology and Biochemistry* 32: 489-496.

COOKE RC, WHIPPS JM (1993). *Ecophysiolgy of fungi.* Oxford, UK: Blackwell Scientific Pubblications.

CORNELISSEN JHC, AERTS R, CERABOLINI B, WERGER MJA, van der Heijeden MGA. 2001. Carbon cycling traits of plant species are linked with mycorrhizal strategy. *Oecologia* 129: 611-619.

COURTY PE, PRITSCH K, SCHLOTER M, HARTMANN A, GARBAYE J (2005). Activity profiling of ectomycorrhiza communities in two forest soils using multiple enzymatic tests. *New Phytologist* 167: 309-319.

COURTY PE, POUYSEGUR R, BUEE M, GARBAYE J (2006). Laccase and phosphatase activities of the dominant ectomycorrhizal types in a lowland oak forest. *Soil Biology and Biochemistry* 38: 1219-1222.

COX J, KAUTZ R, MACLAUGHLIN M, GILBERT T (1994). *Closing the gaps in Florida's wildlife habitat conservation system.* Rep., Off. Environ. Serv., Fla. Game Fresh Water Fish Comm., Tallahassee.

DAHLBERG A (2001). Community ecology of ectomycorrhizal fungi: an advancing interdisciplinary field. *New Phytologist* 150: 555-562.

DANIELSON RM (1984). Ectomycorrhizal associations in jack pine stands in north-eastern Alberta. *Canadian Journal of Botany* 62: 932–939.

DANIELSON RM, VISSER S (1989). Effects of forest soils acidification on ectomycorrhizal and vescicular-arbuscular mycorrhizal development. *New Phytologist* 112: 41-47.

DAWSON TE, MAMBELLI S, PLAMBOECK A, TEMPLER PH, Tu KP (2002). Stable isotopes in plant ecology. *Annual Review of Ecology and Systematic* 33: 507-559.

de BARY A (1887). Comparative morphology and biology of the fungi, mycetozoa and bacteria. Clarendon Press, Oxford.

DE ROMAN M, DE MIGUEL AM (2005). Post- fire, seasonal and annual dynamics of the ectomycorrhiza community of *Quercus ilex* L. forest over a 3 years. *Mycorrhiza* 15 (6): 471-482.

DELWICHE CC, ZINKE PJ, JOHNSON CM, VIRGINIA RA (1978). Nitrogen isotope distribution as a presumptive indicator of nitrogen fixation. *Botanical Gazette* 140: S65-S69.

DICKIE IA, XU B, KOIDE RT (2002). Vertical niche differentiation of ectomycorrhizal hyphae in soil as shown by T-RFLP analysis. *New Phytologist* 156: 527-535.

DICKIE IA, XU B, KOIDE RT (2003). Vertical niche differentiation of ectomycorrhizal hyphae in soil as shown by T-RFLP analysis. *New Phytologist* 156: 527-535.

DIGHTON J, MORALE BONILLA AS, JIMINEZ-NUNEZ RA, MARTINEZ N (2000). Determinants of leaf litter patchiness in mixed species New Jersey pine barrens forest and its possibile influence on soil and soil biota. *Biology and Fertility of Soils* 31: 288-293.

DIGHTON J, SKEFFINGTON RA (1987). Effects of artificial acid precipitation on the mycorrhizas of Scots pine seedlings. *New Phytologist* 107: 191-202.

DUFR'ENE M, LEGENDRE P (1997). Species assemblages and indicator species: the need for a flexible asymmetrical approach. *Ecological Monographs* 67:345–66.

EBERHRADT U, WALTER L, KOTTKE I (1999). Molecular and morphologial discrimination between *Tylospora fibrillosa* adn *Tylospora asterophora* mycorrhizae. *Canadian Journal of Botany* 75: 1323-1335.

EDWARDS GS, KELLY JM (1992). Ecto-mycorrhizal colonisation of loblolly-pine seedlings during 3 growing seasons in response to ozone, acid precipitation and soil Mg status. *Environment and Pollution* 76: 71-77.

EGGER KN (1995). Molecular analysis of ectomycorrhizal fungal communities. *Canadian Journal of Botany* 73: 1415–1422.

EGGER KN, HIBBETT DS (2004). The evolutionary implications of exploitation in mycorrhizas. Canadian Journal of Botany-Revue Canadienne de Botanique 82: 1110-1121.

EGLI S (1981). Die Mykorrhiza und ihre vertikale Verteilung in Eichenbaständen. *Schweizerischen Zeitschrift für Forstwesen* 132: 345-353.

ERLAND S, SÖDERSTRÖM B (1991). Effects of lime and ash treatments on ecto-mycorrhizal infection on *Pinus sylvestris* L. seedlings planted in a pine forest. *Scandinavian Journal of Forest Research* 6: 519-526.

ERLAND S, TAYLOR AFS (1999). Resupinate Ecto-mycorrhizal Fungal Genera. In: Cairney JWG, Chambers SM (eds) Ecto-mycorrhizal fungi: key genera in profile. Springer, Berlin Heidelberg New York, pp.347-363.

ERLAND S, TAYLOR AFS (2002). Diversity of Ectomycorrhizal Fungal Communites in Relation to the Abiotic Environment. In: *Ecological Studies* Vol. 157 M.G.A. Van der Heijden, I Sanders (Eds.) Mycorrhizal Ecology . Springer-Verlag Berlin Heidelberg.

FASSI B, de VECCHI E (1962). Ricerche sulle micorrize ectotrofiche del *Pino strobo* in vivaio.I. Descrizione di alcune forme più diffuse in Piemonte. *Allinoia* 8: 133-151.

FIORE-DONNO AM, MARTIN F (2001). Populations of ectomycorrhizal *Laccaria amethystina* and *Xerocomus* spp. show contrasting colonization patterns in a mixed forest. *New Phytologist* 152: 533–542.

FITTER AH, MOYERSOEN B (1996). Evolutionary trends in root-microbe symbioses. *Philosophical Transactions of theRoyal Society of London. Series B-Biological Sciences* 351: 1367-1375.

FLEMING LV (1985). Experimental study of sequences of ectomycorrhizal fungi on birch (*Betula* sp.) seedlings root systems. *Soil Biology and Biochemistry* 17: 591-600.

FRANK A (1877). Über die biologischen Verhältnisse des Thallus einiger Krustenflechten. Beiträge zur Biologie der Pflanzen 2: 123-200.

FRANSSON PMA, TAYLOR AFS, FINLAY RD (2000). Effects of continuous optimal fertilisation upon a Norway spruce ectomycorrhizal community. *Tree Physiology* 20: 599-606.

FRANSSON PMA, TAYLOR AFS & FYNLAY RD (2001). Elevated atmospheric CO_2 alters root symbiont community structure in forest trees. *New Phytologist* 152: 431-442.

GARDES M, BRUNS TD (1996). Community structure of ectomycorrhizal fungi in a Pinus muricata forest: Above- and below-ground views. *Canadian Journal of Botany* 74: 1572-1583.

GEBAUER G, DIETRICH P (1993). Nitrogen Isotope Ratios in Different Compartments of a Mixed Stand of Spruce, Larch and Beech Trees and of Understorey Vegetation Including Fungi. *Isompenpraxis Environmental Health Studies* 29: 35-44.

GEBAUER G, MAYER M (2003). ^{15}N and ^{13}C natural abundance of autotrophic and mycohertotrophic orchids provides insight into nitrogen and carbon gain from fungal associations.

New Phytologist 160:209-223.

GEBHARDT S, NEUBERT J, WÖLLECKE B, MÜNZENBERGER B, HÜTTL RF (2007). Ectomycorrhiza communities of red oak (*Quercus rubra* L.) of different age in the Lusatian lignite mining distict, East Germany. *Mycorrhiza* 17: 279-290.

GEHRING CA, THEIMER TC, WHITHAM TG, KEIM P (1998). Ectomycorrhizal fungal communiy structure of pinyon pine growing in two environmental extremes. *Ecology* 79: 1562-1572.

GENNEY DR, ANDERSONO IC, ALEXANDER IJ (2006). Fine-scale distribution of pine ectomycorrhizas and their extramatrical mycelium. *New Phytologist* 170: 381-390.

GLEIXNER G, DANIER H-J, WERNER RA, SCMIDT H-L (1993). Correlations between the ^{13}C content of primary and secondary plant products in different cell compartments and that in decomposing basidiomycetes. *Plant Physiology* 102: 1287-1290.

GODBOLD DL, BERNSTON GM (1997). Elevated atmospheric CO_2 concetration changes ectomycorrhizal morphotype assemblages in Betula papyrifera. *New Phytologist* 17:347-350.

GODBOlD DL, BERNTSON GM, BAZZAZ FA (1997). Growth and mycorrhizal colonization of three North American tree species under elevated atmospheric CO_2. *New Phytologist* 137: 433-440.

GÖTTSCHE D (1972). Mitteilungen der Bundesforschungsamt für Forst – und Holzwirtschaft. Verteilung von Feinwurzeln und Mykorrhizen im Bodenprofil eines Buchen – und Fichtenbestandes im Solling.

GOODMAN DM, TROFYMOW JA (1998). Distribution of ectomycorrhizas in micro-habitats in mature and old-growth stands of Douglas-fir sotheastern Vancouver Island. *Soil Biology and Biochemistry* 30: 2127-2138.

GREBENC T, KRAIGHER H (2007).Types of Ectomycorrhiza of Mature Beech and Spruce at Ozone-Fumigated and Control Forest Plots. *Environmental Monitoring and Assessment* 128: 45-59.

GROGAN P, BAAR J, BRUNS TD (2000). Below-ground ectomycorrhizal community structure in a recently burned bishop pine forest. *Journal of Ecology* 88: 1051–1062.

GRYTA H, DEBAUD J-C, EFFOSSE A, GAY G, MARMEISSE R (1997). Fine-scale structure of populations of the ectomycorrhizal fungus *Hebeloma cylindorsporum* in coastal sand dune forest ecosystems. *Molecular Ecology* 6: 353–364.

GUIDOT A, DEBAUD JC, EFFOSSE A, MARMEISSE R (2003). Below-ground distribution and persistence of an ectomycorrhizal fungus. *New Phytologist* 161: 539-547.

GUIDOT A, DEBAUD JC, MARMEISSE R (2001). Correspondence between genet diversity and spatial distribution of above- and below-ground populationsof the ectomycorrhizal fungus *Hebeloma cylindrosporum. Molecular Ecology* 10: 1121–1131.

HARLEY JL (1948). Mycorrhiza and soil ecology. *Biological Review* 23: 127-158.

HARLEY JLF, HARLEY EL (1987). A check-list of mycorrhiza in the british flora. *New Phytologist* 105: 1-102.

HARLEY JL, WAID JS (1955). A method of studying active mycelia on livings roots and other surfaces in the soil. *Transaction of the British Mycological Society* 38: 104-118.

HARTLEY J, CARINEY JWG, FREESTONE O, WOODS C, MEHARG AA (1999). The effects of multiple metal contamination on ecto-mycorrhizal Scots pine (*Pinus sylvestris*) seedlings. *Environment and Pollution* 106: 413-424.

HAUG I, WEISS M, HOMEIER J, OBERWINKLER F, KOTTKE I (2005). Russulaceae and Thelephoraceae form ectomycorrhizas with members of the Nyctaginaceae (Caryophyllales) in the tropical mountain rain forest of southern Ecuador. *New Phytologist* 165: 923-936.

HARVEY AE, JURGENSEN MF, LARSEN MJ, GRAHAM RT (1987). Relationships among soil microsite, ectomycorrhizae and natural conifer regeneration of old-growth forests in western Montana. *Canadian Journal of Forest Research*: 17: 58-62.

HEINONSALO J (2004).The effects of forestry practices on ectomycorrhizal fungal communities and seedling establishment. Integrated studies on biodiversity, podzol profile, clear-cut logging impacts and seedling inoculation. Academic Dissertation in General Microbiology Faculty of Biosciences of the University of Helsinki.

HEINONSALO J, HURME KR, SEN R (2004). Recent C-14-labelled assimilate allocation to Scots pine seedling root and mycorrhizosphere compartments developed on reconstructed podzol humus, E- and B-mineral horizons. *Plant and Soil* 259: 111-121.

HEINONSALO J, JØRGENSEN K, SEN R (2001). Microcosm-based analyses of Scots pine seedling growth, ectomycorrhizal fungal communiy structure and bacterial carbon utilization profiles in boreal forest humus and underlying illuvial mineral horizons. FEMS *Microbiology Ecology* 36: 73-84.

HENN MR, CHAPELA IH (2001). Ecophysiology of C-13 and N-15 isotopic fractionation in forest fungi and the roots of the saprotrophic-mycorrhizal divide. *Oecologia* 128: 480-487.

HIBBETT DS, GILBERT LB, DONOHUGUE MJ (2000).Evolutionary instability of ectomycorrhizal symbioses inbasidiomycetes. *Nature* 407: 506-508.

HOBBIE EA (2006). Carbon allocation to ectomycorrhizal fungi correlates with belowground allocation in culture studies. *Ecology* 87: 563-569.

HOBBIE EA, WEBER NS, TRAPPE JM (2001). Mycorrhizal vs saprotrophic status of fungi: the isotopic evidence. *New Phytologist* 150: 601-610.

HÖGBERG MN, HÖGBERG P (2002). Extramatrical ectomycorrhizal mycelium contributes one-third of microbial biomass and produces, together with associated roots, half the dissolved organic carbon in a forest soil. *New Phytologist* 154:791–795.

HÖGBERG P, HÖGBOM L, SCHINKEL H, HÖRGBERG M, JOHANNISSON C, WALLMARK H (1996). 15N abundance of surface soils, roots, and mycorrhizas in profiles of European forest soils. *Oecologia* 108: 297-214.

HÖGBERG P, PLAMBOECK AH, TAYLOR AFS, FRANSSON PMA (1999). Natural 13C abundance reveals trophic status of fungi and host-origin of carbon in mycorrhizal fungi in mixed forests. *Proceedings of the National Academy of Sciences of the United States of America* 96: 8534-8539.

HORTON T, BRUNS TD (2001). The molecular revolution in ectomycorrhizal ecology: Peeking into the black-box. *Molecular Ecology* 10: 1855-1871.

HORTON T, BRUNS TD, PARKER VT (1999). Ectomycorrhizal fungi associated with *Arctostaphylos* contribute to *Pseudotsuga menziesii* establishment. *Canadian Journal of Botany* 77: 93-102.

INEICHEN K, WIEMKEN V, WIEMKEN A (1995). Shoots, roots and ectomycorrhiza formation of pine seedlings at elevated atmospheric carbon dioxide. *Plant Cell and Environment* 18: 703-707.

ISHIDA TA, KAZUHIDE N, HOGETSU T (2007). Host effects on ectomycorrhizal fungal communities: insight from eight host species in mixed conifer-broadleaf forests. *New Phytologist* 174: 430-440.

IZZO A, AGBOWO J, BRUNS TD (2004). Detection of plot-level changes in ectomycorrhizal communities across year in an old-growth mixed conifer forest. *New Phytologist* 166: 619-630.

IZZO A, AGBOWO J, BRUNS TD (2005). Detection of plot-level changes in ectomycorrhizal communities across year in an old-growth mixed-conifer forest. *New Phytologist* 166: 619–629.

IZZOA, CANRIGHT M, BRUNS TD (2006).The effects of heat treatments on ectomycorrhizal resistant propagules and their ability to colonize bioassay seedlings. *Mycological Research* 110: 196-202.

JACKSON RB, CANADELL J, EHLERINGER JR, MOONEY HA, SALA OE, SCHULZE ED (1996). A global analysis of root distribution for terrestrial biomes. *Oecologia* 108: 389-411.

JOHNSON BC, GRAHAM JH, SMITH FA (1997). Functioning of mycorrhizal associations along the mutualism-parasitism continuum. *New Phytologist* 135:575-585.

JOHNSON D, MARLEEN IJDO, GENNEY DR, ANDERSON IC, ALEXANDE IJ (2005). How do plants regulate the function, community structure, and diversity of mycorrhizal fungi? *Journal of Experimental Botany* (56) 417: 1751–1760.

JONES MD, DURALL DM, TINKER PB (2001). Fluxes of carbon and phosphorus between symbionts in willow ectomycorrhizas and their changes with time. *New Phytologist* 119: 99-106.

JONSSON L, DAHLBERG A, NILSSON M-C, ZACKRISSON O, KÅRÉN O (1999). Ectomycorrhizal fungal communities in late-successional Swedish boreal forest, and their composition following wildfire. *Molecular Ecology* 8: 205-212.

JONSSON LM, NILSSON MC, WARDLE DA, ZACKRISSON O (2001). Context dependent effects of ectomycorrhizal species richness on tree seedling productivity. *Oikos* 93: 353-364.

KÅRÉN O, NYLUND J-E (1996). Effects of N-free fertilization on ecto-mycorrhiza community structure in Norway spruce stands in southern Sweden. *Plant Soil* 181: 295-305.

KASURINEN A, KEINANEN MM, KAIPAINEN S, NILSSON LO, VAPAAVUORI E, KONTRO MH, HOLOPAINEN T (2005). Belowground responses of silver birch trees exposed to elevated CO_2 and O-3 levels during three growing seasons. *Global Change Biology* 11:1167-1179.

KOIDE R, T XU B, SHARDA J, LEKBERG Y, OSTIGUY N (2004). Evidence of species interactions within an ectomycorrhizal fungal community. *New Phytologist* (on line).

KERNAGHAN G, WIDDEN P, BERGERON Y, LE'GARE' S, PARE' D (2003). Biotic and abiotic factors affecting ectomycorrhizal diversity in boreal mixed-woods. *Oikos* 102: 497-504.

KIELLAND K (1994). Amino acid absorption by arctic plants: Implications for plant nutrition and nitrogen cycling. *Ecology* 75: 2373-2383.

KINGSTON JC, BIRKS HJB, UUTALA AJ, CUMMING BF, SMOL JP (1992). Assessing trends in fishery resources and lake water aluminium from paleolimnological analyses of siliceous algae. *Canadian Journal Fisheries Aquatic Sciences.* 49:116–27.

KOIDE RT, COURTY PE, GARBAYE J (2007). Research perspectives on functional diversity in ectomycorrhizal fungi. *New Phytologist* 174: 240-243.

KOIDE TR, SHUMWAY DL, XU B, SHARDA JN (2007). On temporal partioning of a community of ectomycorrhizal fungi. *New Phytologist* 174: 420-429.

KOIDE RT, XU B, SHARDA J (2004). Contrasting below-ground views of an ectomycorrhizal fungal community. *New Phytologist* 166: 251-262.

KOIDE R, XU B, SHARDA J, LEKBERG Y, OSTIGUY N (2005). Evidence of species interactions within an ectomycorrhizal fungal community. *New Phytologist* 165: 305-316.

KÕLJALG U, DAHLBERG A, TAYLOR AFS, LARSSON E, HALLENBERG N, STENLID J, LARSSON K-H, FRANSSON PM, KARÉN O, JONSSON L (2000). Diversity and abundance of resupinate thelephoroid fungi as ectomycorrhizal symbionts in Swedish boreal forests. *Molecular Ecology* 9: 1985-1996.

KÕLJALG U, LARSSON K-H, ABARENKOV K, NILSSON RH, ALEXANDER IJ, EBERHARDT U, ERLAND S, HOILAND K, KJOLLER R, LARSSON E, PENNANEN T, SEN R, TAYLOR AFS,

TEDERSOO L, VRALSTAD T, URSING BM. (2005) UNITE: a database providing web-based methods for the molecular identification of ectomycorrhizal fungi. New Phytologist 166: 1063-1068.

KOTTKE I, BEITER A, WEISS M, HAUG I, OBERWINKLER F, NEBEL M (2003). Heterobasidiomycetes form symbiotic associations with hepatics: Jungermanniales have sebacinoid mycobionts while *Aneura pinguis* (Metzgeriales) is associated with a *Tulasnella* species. *Mycological Research* 107: 957-968.

KOTTKE I, OBERWINKLER F (1987). The cellular structure of the Hartig net: coenocytic and transfer cell-like organisation. *Nordic Journal of Botany* 7: 85-95.

KOVACS G, PAUSCH M, URBAN A (2000). Diversity of Ectomycorhizal Morphotypes and Oak Decline. *Phyton* (Horn, Austria), 40(4), 109–116.

KRAIGHER H (1999). Diversity of types of Ectomycorrhizae on Norway spruce in Slovenia. *Phyton* (Horn, Austria) 39: 199–202.

KRAIGHER H, AL SAYEGH PETKOVŠEK S, GRENBEC T, SIMONČIČ P (2006). Types of ectomycorrhiza as Pollution Stress Indicators: Case Studies in Slovenia. *Environmental Monitoring Assessment* DOI 10.1007/s10661-006-9413-4.

KRAIGHER H, BATIČ F, AGERER R (1996). Types of ectomycorrhizae and mycobioindication of forest site pollution. *Phyton (Horn, Austria)* 36:115-120.

KUITERS AT (1990). Role of phenolic substances from decomposing forest litter in plant-soil interaction. *Acta Botanica Neerlandica* 39:329-348.

KURTZ JC, JACKSON LE, FISHER WS (2001). Strategies for evaluating indicators based on guidelines from the Environmental Protection Agency's Office of Research and Development. *Ecological Indicators* 1:49–60.

LAIHO O, MIKOLA P (1964). Studies on the effect of some eradicants on mycorrhizal development in forest nurseries. *Acta Forestalia Fennica* 77: 1-34.

LAMBECK RJ (1997). Focal species: a multispecies umbrella for nature conservation. *Conservation Biology* 11:849–56.

LANDRES PB, VERNER J, THOMAS JW (1988). Ecological uses of vertebrate indicator species: a critique. *Conservation Biology* 2:1–10.

LANDEWEERT R, LEEFLANG P, KUYPER TW, HOFFLAND E, ROSLING A, ERNARS K, SMIT E (2003). Molecular identification of ectomycorrhizal mycelium in soil horizons. *Applied and Environmental Microbiology* 69: 327-333.

LARSEN MJ, SMITH JS, McKAY (1997).On *Piloderma bicolor* and the closely related *P. byssinum*, *P. Croceum*, and *P. fallax*. *Mycotaxon* 63: 1-8.

LARSSON KH, LARSSON E, KÕLJALG U (2004). High phylogenetic diversity among corticoid homobasidiomycetes. *Mycological Research* 108: 983-1002.

LAST FT, DIGHTON J, MASON PA (1987). Successions of sheathing mycorrhizal fungi. *Trends in Ecology and Evolution* 2: 157–161.

LEAKE JR, READ DJ (1997). Mycorrhizal fungi in terrestrial ecosystems. In: Wicklow D, Söderström B, eds. *The Mycota IV environmental and microbial relationships*. Berlin, Germany: Springer-Verlag, 281-301.

LEE KJ, KIM YS (1987). Host specificity and distribution of putative ectomycorrhizal fungi in pure stands of twelve tree species in Korea. *Korean Journal of Mycology* 15: 48-69.

LEE LS, ALEXANDER IJ (1996). The dynamics of ectomycorrhizal infection of Shorea leprosula seedlings in Malaysian rain forests. *New Phytologist* 132: 297–305.

LEHTO T (1984). Kalkituksen vaikutus männyn mykoritsoihin. *Folia Forestalia* 609: 1-20.

LEHTO T (1994). Effects of liming and boron fertilization on mycorrhizas of *Picea abies*. *Plant Soil* 163: 65-68.

LEYVAL C, TURNAU K, HASELWANDTER K (1997). Effect of heavy metal pollution on mycorrhizal colonization and function: physiological, ecological and applied aspects. *Mycorrhiza* 7: 139-153.

LILLESKOV EA, FAHEY TJ (1996). Patterns of ecto-mycorrhizal diversity over an atmospheric nitrogen deposition gradient near Kenai, Alaska. In: Szaro TM, Bruns TD (eds) Abstracts of the 1st International Conference on Mycorrhizae: Univ California, Berkeley, 76.

LILLESKOV EA, FAHEY TH, HORTON TR, LOVETT GM (2002). Belowground ectomycorrhizal fungal community change over a nitrogen deposition gradient in Alaska. *Ecology* 83: 104-115.

LILLESKOV EA, FAHEY TJ, LOVETT GM (2001). Ectomycorrhizal fungal aboveground community change over an atmospheric nitrogen deposition gradient. *Ecological Applications* 11: 397-410.

LIPSON DA, SCHADT CW, SCHMIDT SK, MONSON RK (1999). Ectomycorrhizal transfer of amino acid-nitrogen to the alpine sedge *Kobresia myosuroides*. *New Phytologist* 142: 163-167.

MACDOUGAL DT, DUFRENOY J (1946). Criteria of nutritive relations of fungi and seed-plants in mycorrhiza. *Plant Physiology* 21, I.

MAGURRAN AE (1988). Ecological diversity and its measurement. Croom Helm, London.

MAHMOOD S (2000). Ecto-mycorrhizal community structure and function in relation to forest residue harvesting and wood ash applications. Doctoral Thesis, Lund University. ISBN 91-7105-136-8.

MAHMOOD S, FINLAY RD, ERLAND S (1999). Effects of Repeated Harvesting of Forest

Residues on the Ectomycorrhizal Community in a Swedish Spruce Forest. *New Phytologist* 142(3): 577-585.

MAKKONEN K, HELMISSARI HS (1998). Seasonal and yearly variation of fine-root biomass and necromass in a Scot pine (*Pinus sylvestris* L.) stand. *Forest Ecology and Management* 102: 283-290.

MALAJCZUK N, HINGSTON FJ (1981). Ectomycorrhizae associated with Jarrah. *Australian Journal of Botany* 29: 453–462.

MAMOUN M, OLIVIER JM (1993a). Competition between *Tuber melanosporum* and other ectomycorrhizal fungi under two irrigation regimes. I. Competition with *Tuber brumale*. *Plant and Soil* 149: 211–218.

MAMOUN M, OLIVIER JM (1993b). Competition between *Tuber melanosporum* and other ectomycorrhizal fungi under two irrigation regimes. II. *New Phytologist* 165: 305–316.

MANLEY PN, ZIELINSKI WJ, SCHLESINGER MD, MORI SR (2004). Evaluation of a multiple species approach to monitoring species at the ecoregional scale. *Ecological Application* 14: 296–310.

MASON PA, LAST FT, PELHAM J, INGLEBY K (1982). Ecology of some fungi associated with an aging stand of birches (*Betula pendula* and *Betula pubescens*). *Forest Ecology and Management* 4: 19–39.

MASON PA, WILSON J, LAST FT, WALKER C (1983). The concept of succession in relation to the spread of sheathing mycorrhizal fungi on inoculated tree seedlings growing in unsterile soils. *Plant and Soil* 71: 47–256.

MASSICOTTE HB, MELVILLE LH, PETERSON RL, LUOMA DL (1998). Anatomical aspects of field ectomycorrhizas on *Polygonum viviparum* (Polygonaceae) and *Kobresia bellardii* (Cyperaceae). *Mycorrhiza* 7: 287-292.

MASSICOTTE HB, MOLINA R, TACKABERRY LE, SMITH JE, AMARANTHUS MP (1999). Diversity and host specificity of ectomycorrhizal fungi retrieved from three adjacent forest sites by five host species. *Canadian Journal of Botany* 77: 1053-1076.

McCUNE B, MEFFORD MJ (1999). PC-ORD.Multivariate Analysis of Ecological Data.Version 5.0. MjM Software, Gleneden Beach, Oregon, U.S.A.

MEYER FH (1962). Die Buchen und Fichtenmykorrhiza in verschiedenen Bodentypen, ihre Beeinflussung durch Minealdünger sowie für die Mykorrhizabildung wichtige Faktoren. Mitteilungen der Budesforschungsanstalt für Forst- und Holzwirtschaft 54: 1-73.

MEYER FH (1973). Distribution of ectomycorrhizae in native and man-made forests. In: *Ectomycorrhizae* (eds G. C. Marks, and T. T. Kozlowski). Academic Press, New York, USA. pp. 79-

105.
MICHELSEN A, SCHMIDT IDK, JONASSON S, QUARMBY C, SLEEP D (1996). Leaf ^{15}N abundance of subarctic plants provides field evidence that ericoid, ectomycorrhizal and non- and arbuscular mycorrhizal species access different sources of soil nitrogen. *Oecologia* 105: 53-63.

MOLINA R, MASSICOTTE H, TRAPPE JM (1992). Specificity phenomena in mycorrhizal symbioses: community-ecological consequences and practical implications. In: Allen MJ, ed. *Mycorrhizal functioning*. New York, NY, USA: Chapman & Hall, 357-423.

MOSCA E (2007). Rapporto tra deperimento della farnia e stato ectomicorrizico: effetto di un trattamento selvicolturale. PhD thesis Università degli Studi di Padova-Université Henri Poincaré Nancy1.

MOSCA E, MONTECCHIO L, SELLA L, GARBAYE J (2007).Short-term effect of removing tree competition on the ectomycorrhizal status of a declining pedunculate oak forest (*Quercus robur* L.). *Forest Ecology and Management* 244: 129-140.

MURAT C, DÍEZ J, LUIS P, DELARUELLE C, DUPRE C, CHEVALIER G, BONFANTE P, MARTIN F (2004). Polymorphism at the ribosomal DNA ITS and its relation to postglacial re-colonization routes of the Perigord truffle *Tuber melanosporum*. *New Phytologist* 164: 401-411.

NANTEL P, NEUMANN P (1992). Ecology of ectomycorrhizal-basidiomycete communities on a local vegetation gradient. *Ecology* 73: 99-117.

NEWTON AC, HAIGH JM (1998). Diversity of ectomycorrhizal fungi in Britain: a test of the species-area relationship, and the role of host specificity. *New Phytologist* 138: 619-627.

NIEMI JG, McDONALD EM (2004).Application of ecological indicators. *Annual Review Ecology Evolotion and Systematics* 35:89–111.

NORBY RJ, O'NEILL EG, HOOD WG, LUXMORE RJ (1987). Carbon allocation, root exudation and mycorrhizal colonization of *Pinus echinata* seedlings grown under CO_2 enrichment. *Tree Physiology* 3: 203-210.

O'CONNOR RJ, WALLS TE, HUGHES RM (2000).Using multiple taxonomic groups to index the ecological condition of lakes. *Environmental Monitoring and Assessment* 61:207–2.

O'DELL TE, AMMIRATI JF, SCHREINER EG (1999). Species richness and abundance of ectomycorrhizal basidiomycete sporocarps on a moisture gradient in the *Tsuga heterophylla* zone. *Canadian Journal Botany* 77: 1699-1711.

OLIVIER J-M, MAMOUN M (1994). Compétitions entre symbiotes sur jeunes noisetiers truffiers. *Acta Botanica Gallica* 141: 559-563.

OLSSON PA, MÜNZENBERGER B, MAHMOOD S, ERLAND S (2000). Molecular and anatomical evidence for a three-way association between *Pinus sylvestris* and the ectomycorrhizal fungi *Suillus bovinus* and *Gomphidius roseus*. *Mycological Research* 104: 1372–1378.

PENUELAS J, BOADA M (2003). A global change-induced biome shift in the Montseny mountains (NE Spain). *Global Change Biology* 9:131–140.

PERSSON H (1980). Spatial distribution of fine-root growth, mortality and decomposition in a young Scots pine stand in Central Sweden. *Oikos* 34: 77-97.

PERSSON H, AHLSTRÖM K (1994). The effects of alkalizing compounds on fine-root growth in a Norway spruce stand in southwest Sweden. *Journal Environmental Science Health* 29: 803-820.

POST DM (2002). Using stable isotopes to estimate trophic position: Models, methods and assumptions. *Ecology* 83: 703-718.

QIU Z, CHEPELLKA AH, SOMERS GL, LOCKABY BG, MELDAHL RS (1993). Effects on Ozone and simulated acid precipitation on ecto-mycorrhizal formation on loblolly pine seedlings. *Environmental Experimental Botany* 33: 423-431.

RAIDL S (1997). Studien zur Ontogenie an Rhizomorphen von Ektomykorrhizen. Brbl Mycol 169:1–184.

RAO CS, SHARMA GD, SHUKLA AK (1997). Distribution of ectomycorrhizal fungi in pure stands of different age groups of Pinus kesiya. *Canadian Journal of Microbiology* 43: 85–91.

READ DJ (1991). Mycorrhizas in ecosystems. *Experimenta* 47: 376-391.

READ DJ, LEAKE JR, PEREZ-MORENO J (2004). Mycorrhizal fungi as drivers of ecosystem processes in heathland and boreal forest biomes. *Canadian Journal of Botany* 82: 1243-1263.

READ DJ, PEREZ-MORENO J (2003). Mycorrhizas and nutrient cycling in ecosystems – a journey towards relevance? *New Phytologist* 157: 475-492.

REY A, JARVIS PG (1997). Growth response of young birch trees (*Betula pendula*l Roth.) after four and a half years of CO_2 exposure. *Annals of Botany* 80: 809-816.

REYNOLDSON TB, NORRIS RH, RESH VH, DAY KE, ROSENBERG DM (1997). The reference condition: a comparison of multimetric and multivariate approaches to assess waterquality impairment using benthic macroinvertebrates. Journal of the North American Benthological Society 16: 833–52.

RILLIG MC, KATHLEEN KT, MICHAEL FA (2002). Global Change and Mycorrhizal Fungi. In: *Ecological Studies* Vol. 157 M.G.A. Van der Heijden, I Sanders (Eds.) Mycorrhizal Ecology. Springer-Verlag Berlin Heidelberg.

RILLIG MC, WRIGHT SF, ALLEN MF, FIELD CB (1999). Long-term CO_2 elevation affects soil structure of natural ecosystems. *Nature* 400: 628.

ROSLING A, LANDEWEERT R, LINDAHL BD, LARSSON KH, KUYPER TW, TAYLOR AFS, FINLAY RD (2003). Vertical distribution of ectomycorrhizal fungal taxa in a podzol soil profile.

New Phytologist 159: 775-783.

ROTH DR, FAHEY TJ (1998). The effects of acid precipitation and ozone on the ecto-mycorrhiza of red spruce saplings. *Water Air Soil Pollution* 103: 263-276.

ROUHIER H, READ DJ (1998). Plant and fungal responses to elevated atomospheric carbon dioxide, in mycorrhizal seedlings of *Pinus sylvestris*. *Environmental Experimental Botany* 40(3): 237-246.

ROUHIER H, READ DJ (1999). Plant and fungal responses to elevated CO_2 in mycorrhizal seedlings of Betula pendula. *Environmetal Experimental Botany* 42: 231-241.

RÜHLING Å, SÖDERSTRÖM B (1990). Changes in fruitbody production of mycorrhizal and litter decomposing macromycetes in heavy metal polluted coniferous forests in north Sweden. *Water Air Soil Pollution* 49: 375-387.

RUNION GB, MITCHELL RJ, ROGERS HH, PRIOR SA, COUNTS TK (1997). Effects of nitrogen and water limitation and elevated atmospheric CO2 on ecto-mycorrhiza of longleaf pine. *New Phytologist* 137: 681-689.

SAGARA N (1995). Association of ectomycorrhizal fungi with decomposed animal wastes in forest habitats a cleaning symbiosis? *Canadian Journal of Botany* 73: 1423–1433.

SAKAKIBARA SM, JONES MD, GILLESPIE M, HAGERMAN SM, FOREST ME, SIMARD SW, DURALL DM (2002). A comparison of ectomycorrhiza identification based on morphotyping and PCR-RFLP analysis. *Mycological Research* 106: 868-878.

SCATTOLIN L (2007). Variations of the ectomycorrhizal community in high mountain Norway spruce stands and correlations with the main pedoclimatic factors. PhD thesis - Università degli Studi di Padova and Ludwig-Maximilians- Universität München.

SCATTOLIN L, MONTECCHIO L, AGERER R (2007). The Ectomycorrhizal community structure in high mountain Norway spruce stands. Trees. DOI 10.1007/s00468-007-0164-9. In press.

SHANNON CE,WEAVER W (1949). *The Mathematical Theory of Communication*. Urbana: Univ. Illinois Press.

SHI L, GUTTENBERGER M, KOTTKE I, HAMPP R (2002). The effect of drought on mycorrhizas of beech (*Fagus sylvatica* L.): changes in community structure, and the content of carbohydrates and nitrogen storage bodies of the fungi. *Mycorrhiza* 12: 303–311.

SIMARD SW, PERRY DA, JONES MD, MYROLD DD, DURALL DM, MOLINA R (1997). Net transfer of carbon between ectomycorrhizal tree species in the field. *Nature* 388: 579-582.

SIMON TP, ED (2003). *Biological Response Signatures: Indicator Patterns Using Aquatic Communities*. Boca Raton, FL: CRC

SMEETS E, WETERINGS R 1999. Environmental indicators: typology and overview. *Tech. Rep. 25*, Eur. Environ. Agency, Copenhagen, Den. http://reports.eea.eu.int:80/TEC25/en/ tech 25 text.pdf

SMITH SE, READ DJ (1997). Mycorrhizal symbiosis. San Diego, CA, USA: Academic Press.

STADDON PL, HEINEMEYER A, FITTER AH (2002). Mycorrhizas and global environmental change: research at different scales. *Plant and Soil* 244: 253–261.

STARK JM, HART SC (1997). High rates of nitrification and nitrate turnover in undisturbed coniferous forests. *Nature* 385: 61-64.

STENSTRÖM E (1991) The effects of flooding on the formation of ectomycorrhizae in *Pinus sylvestris* seedlings. *Plant Soil* 131:247-250.

SYLVIA DM, JARSTFER AG (1997). Distribution of mycorrhiza on competing pines and weeds in a southern pine plantation. *Soil Science Society of America Journal* 61:139-144.

TAYLOR AFS (2002) Fungal diversity in ectomycorrhizal communities: sampling effort and species detection. *Plant and Soil* 244: 19–28.

TAYLOR AFS, ALEXANDER IJ (2005). The ectomycorrhizal symbiosis: life in the real world *Mycologist*, Volume 19, Part 3 103-112.

TAYLOR DL, BRUNS TD, SZARO TM, HODGES SA (2003). Divergence in mycorrhizal specialization within *Hexalectris spicata* (Orchidaceae), a nonphotosynthetic desert orchid. *American Journal of Botany* 90: 1168-1179.

TAYLOR AFS, FRANSSON PM, HÖGBERG P, HÖGBERG MN, PLAMBOECK AH (2003). Species level pattern in ^{13}C and ^{15}N abundance of ectomycorrhizal and saprotrophic fungal sporocarps. *New Phytologist* 159: 757-774.

TAYLOR AFS, GEBAUER G, READ DJ (2004). Uptake of nitrogen and carbon from double-labelled (N-15 and C-13) glycine by mycorrhizal pine seedlings. *New Phytologist* 164: 383-388.

TAYLOR AFS, HÖGBOM L, HÖGBERG M, LYON AJE, NASHOLM T (1997). Natural 15N abundance in fruit bodies of ectomycorrhizal fungi from boreal forests. *New Phytologist* 136: 713-720.

TAYLOR AFS, MARTIN F, READ DJ (2000). Fungal diversity in ectomycorrhizal communities of Norway spruce (*Picea abies* (L.) Karst.) and Beech (*Fagus sylvatica* L.) along north-south transects in Europe. In: Schulze ED, ed. Carbon and nitrogen cycling in European forest ecosystems. Berlin: Springer, 343-365.

TEDERSOO L, SUVI T, BEAVER K, KÕLJALG U (2007). Ectomycorrhizal fungi of the Seychelles: diversity patterns and host shifts from the native *Vateriopsis seychellarum* (Dipterocarpaceae) and *Intsia bijuga*(Caesalpiniaceae) to the introduced *Eucalyptus robusta* (Myrtaceae), but not *Pinus caribea* (Pinaceae). *New Phytologist* ***doi***: 10.1111/j.1469-

8137.2007.02104.x

TER BRAAK CJF, ŠMILAUER P (eds) (2002). CANOCO Reference manual and CanoDraw for Windows User's guide: Software for Canonical Community Ordination (version 4.5). Microcomputer Power, Ithaca, NY, USA.

TERMOSHUIZEN AJ (1991). Succession of mycorrhizal fungi in stands of Pinus sylvestris in the Netherlands. In: Termoshouizen AJ (Ed.), Decline of Carpophores of Mycorrhizal Fungi in Stands of *Pinus sylvestris*. PhD Thesis. University, The Netherlands, pp.41-50.

TINGEY DT, PHILLIPS DL, JOHSON MG, STROM MJ, BALL JT (1997). Effects of elevated CO$_2$ and N fertilization on fine root dynamics and fungal growth in seedling *Pinus ponderosa*. *Environmental Experimental Botany* 37: 73-83.

TISDALL JM, OADES JM (1982). Organinc matter and water-stable aggregats in soils. Journal of Soil Science 33: 141-163.

TOLJANDER JF, EBERHARDT U, TOLJANDER YK, PAUL LR, TAYLOR AFS (2006). Species composition of an ectomycorrhizal fungal community along a local nutrient gradient in a boreal forest. *New Phytologist* 170: 873-884.

TRESEDER KK (2005). Unearthing ectomycorrhizal dynamics. *New Phytologist* 166: 358-359.

TRESEDER KK, ALLEN MF (2000). Mycorrhizal fungi have a potential role in soil carbon storage under elevated CO$_2$ and nitrogen deposition. *New Phytologist* 147: 189–200.

TRUDELL SA, RYGIEWICZ PT, EDMONDS RL (2004). Patterns of nitrogen and carbon stable isotope ratios in macrofungi, plants and soils in two old-growth conifer forests. *New Phytologist* 164: 317–335.

UNESTAM T (1991). Water repellency, mat formation, and leaf-stimulated growth of some ectomycorrhizal fungi. *Mycorrhiza* 1: 13-20.

UNESTAM T, STENSTRÖM E (1989). A method for observing and manipulating roots and root-associated fungi on plants growing in nonsterile substrates. *Scandinavian Journal Forest Research* 4:51-58.

UNESTAM T, SUN YP (1995). Extramatrical structures of hydrophobic and hydrophilic ectomycorrhizal fungi. *Mycorrhiza* 5: 301-311.

URBAN A, WEISS M, BAUER R (2003). Ectomycorrhizae involving sebacinoid fungi and their analogs. *Journal of theLinnaean Society of London, Botany* 13: 31-42.

US EPA (2000). Evaluation guidelines for ecological indicators. *EPA/620/R-99/005*, Research Triangle Park, NC.109 pp. http://www.ecosystemindicators.org/wg/publication/Jackson_Kurtz_Fisher.pdf

US EPA (2002a). Evaluation guidelines for ecological indicators. *EPA/620/R-99/005*, Research

Triangle Park, NC.109 pp. http://www.ecosystemindicators.org/wg/publication/Jack son Kurtz Fisher.pdf

US EPA (2002b). A SAB report: a framework for assessing and reporting on ecological condition. *EPA-SAB-EPEC-02–009*, Washington, DC. 142 pp. http://www.epa. gov/sab/pdf/epec02009.pdf

VAN DER HEIJDEN M, KLIRONOMOS J, URSIC M ET AL (1998). Mycorrhizal fungal diversity determines plant biodiversity, ecosystem variability and productivity. – *Nature* 396: 69–72.

VISSER S (1995). Ectomycorrhizal fungal succession in Jack pine stands following wildfire. *New Phytologist* 129: 389–401.

VRÅLSTAD T, FOSSHEIM T, SCHUMACHER T (2000). *Piceirhiza bicolorata* – the ectomycorrhizal expression of the Hymenoscyphus ericae aggregate? *New Phytologist* 145: 549-563.

VRÅLSTAD T, SCHUMACHER T, TAYLOR AFS (2002). Mycorrhizal synthesis between fungal strains of the *Hymenoscyphus ericae* aggregate and potential ectomycorrhizal and ericoid hosts. *New Phytologist* 153: 143-152.

WALKER RF, JOHNSON DW, GEISINGER DR, BALL JT (1998). Growth and ectomycorrhizal colonization of ponderosa pine seedlings supplied different levels of atmospheric CO_2 and soil N and P. *Forest Ecology and Management* 109 (1-3): 9-20.

WALLANDER H (2006). External mycorrhizal mycelia – the importance of quantifiation in natural ecosystems. *New Phytologist* 171: 240-242.

WALLANDER H, MAHMOOD S, HAGERBERG D, JOHANSSON L, PALLON J (2003). Elemental composition of ectomycorrhizal mycelia identified by PCR-RFLP analysis and growth in contact with apatite or wood ash in forest soil. FEMS *Microbiological Ecology* 44: 55-65.

WALLANDER H, NILSSON LO, HAGERBERG D, BAATH E (2001). Estimation of the biomass and seasonal growth of external mycelium of ectomycorrhizal fungi in the field. *New Phytologist* 171: 240-242.

WALLENDA T, KOTTKE I (1998). Nitrogen deposition and ectomycorrhizas. *New Phytologist* 139: 169-187.

WEISS M, SELOSSE MA, REXER KH, URBAN A, OBERWINKLER F (2004). Sebacinales: a hitherto overlooked cosm of heterobasidiomycetes with a broad mycorrhizal potential. *Mycological Research* 108: 1003-1010.

WENGER KF (1955). Light and mycorrhiza development. *Ecology* 36: 518-520.

WILSON, WARREN J (1951). Micro-organisms in the rhizosphere of beech. *D. Phil. Thesis, Oxford University* (typescript).

WU B, NARA K, HOGETSU T (1999). Competition between ectomycorrhizal fungi colonizing

Pinus densiflora. Mycorrhiza 9: 151–159.

YOUNG IM, BLANCHART E, CHENU C, DANGERFIELD M, FRAGOSO C, GRIMALDI M, INGRAM J, MONROZIER L (1998). The interaction of soil biota and soil structure under global change. *Global Change Biology* 4: 703-712.

ZHOU Z, HOGETSU T (2002). Subterranean community structure of ectomycorrhizal fungi under Suillus grevillei sporocarps in a Larix kaempferi forest. *New Phytologist* 154: 529-539.

CHAPTER 2

The composition of the ectomycorrhizal community in beech coppices of different age

1. Introduction

The research on ectomycorrhizae (EM) received a great impulse in the last 40 years thanks to the work of many scientists. The first morphological investigation together with the latest ecological, physiological and genetic studies widened the information now available, but the progresses in the anatomical identification of the fungi is a prerequisite of the studies of EM communities (De Roman *et al.*, 2005). The molecular analysis are not sufficient and not always efficient or reliable to classify the species, and morphological in combination with anatomical features have a fundamental role to understand the fungal structure and its different developmental stages on the host. Since the beginning of ectomycorrhizal symbiosis research in the late 19th century, a lot of EM morphotypes have been more or less accurately described, and few authors tried to create a classification systems and to develop identification keys similar to that available for plants and animals, but this was a difficult task (De Roman *et al.*, 2005). The first attempt was made by Dominik (1969) and a few years later Zak (1973) attested that a detailed description of each EM was essential for the identification. Goodman *et al.* (1996–2000) realized concise Descriptions of North American Ectomycorrhizas, and the descriptions published according to this system were more detailed than those in Ingleby *et al.* (1990), but they lacked the level of detail. In the year 1986 Agerer (1986, 1987–2006, 1994, 1999) began to publish guidelines for the systematic descriptions and identification of EM that are widely used nowadays (De Roman *et al.*, 2005). In addition, Agerer created a binomial nomenclature system for those EM described but not yet identified, edited a Colour Atlas of Ectomycorrhizas (Agerer 1987–2006) with photographs of the EM to facilitate identification by comparison, and developed a synoptic key and determine EM (Agerer & Rambold 1998, 2004-2007).

Some 5.000-6.000 fungal species are estimated to be ectomycorrhizal fungi (Molina *et al.* 1992; Taylor & Alexander 2005), but only a small portion of them has been investigated by anatomical studies (Agerer 2006). The most important informative ectomycorrhizal features for the recognition of fungal relationships are:

a) structure of the mantle layer as seen in plan view;

b) structure of rhizomorphs;

c) shape of cystidia;

d) features of emanating hyphae.

In addition all the anatomical features, can be used to characterize EM, in particular those including hyphae (Agerer 2006). Recent investigations about the ecological function of the symbiotic species in the ecosystem, gave the possibility to apply putatively ecologically important features as expressed by their exploration types (Agerer 2001). Up to now only Brand (1991) published a more detailed contribution to the ectomycorrhize on *Fagus sylvatica* L., with 23 descriptions. Here we present the structure of the community discovered in beech coppices of different age in the Province of Trento (Trentino-Südtirol Region in Italy), thanks to three years of research on this topic.

2. Methods

The collected rootlets were carefully cleaned from adehring soil and debris in tap water. Under a stereomicroscope connected to digitals cameras the EM were sorted at first into morphotypes based on colour, occurence and abundance of cystidia, emanating hyphae and rhizomorphs (Agerer 1987-2006, Agerer 1991). Furthermore several root tips of each morphotype per sample were anatomotyped following Agerer (1991). These analyses were completed within 12 days after sampling. Also the available literature was used to classify the anatomotypes (Goodman *et al.* 1996-2000; Agerer 1987-2006; Cairney & Chambers 1999; Brand 1991, Agerer & Rambold 2004-2007; Haug *et al.*, 1994). Subsenquently, EM were classified into exploration types (Agerer 2001) and we noted the hydrophobicity attitude according Unestam (1991).The anotomotypes unidentified by molecular or anatomical tools, were classified by an alphanumerical code (EDMxx). For the other anatomotypes we wrote the name and the alphanumerical code. Sequences taxon categories were assigned as follows: sequence similarity of 100% (= identification to species level) sequence to similarity of 95% to 99% (= identification to genus level) sequence similarity of < 95% (= identification to family or ordinal level).

DNA extraction, amplification and sequencing.—DNA was extracted from organs of ectomycorrhizae after Gardes and Bruns (1993) using a Quiagen DNeasy plant Mini Kit (Quiagen, Hilden, Germany), according to the manufacturer's instructions. PCR amplification was performed for internal transcribed spacers ITS1, ITS2, and for 5.8S region of the nuclear ribosomal DNA, using basidiomycete specific primer pairs ITS1F (5´ cttggtcatttagaggaagtaa 3´) and ITS4B (5´tcctccgcttattgatatgc 3´). PCR amplification was performed using a Ready To Go™ beads (Amersham Pharamacia Biotech., Piscataway, New Jersey), with 20 µm of PCR solution (composed of 120 µm ddH2O, 30 µl buffer, 21,6 µl MgCl, 12 µm ITS1F, 12 µl ITS4B, 30 µl dNTP-Mix and 2,4 µm Taq-Polymerase) and 5 µl extracted DNA. The PCR was programmed as follows: 95 °C for 5 min, [90° 30 sec, 55 °C for 30 sec, 72 °C for 1 min (+ 2 sec for each cycle): 35 cycles], 72°C for 10 min,16 °C infinitely (Tedersoo *et al.*, 2006). Amplified PCR products (2 µl) were run with bromophenol blue (2 µl) on 1% agarose gels for 30 min at 95 W, then stained in ethidium bromide for 10 min and afterward in ddH2O for 1 min. PCR products were then visualised under the UV light. Successful DNA bands were purified using the QIAquick-PCR purification Kit (Qiagen GmbH, Hilden, Germany) according to manufacturer's instructions. DNA sequencing was performed by the sequencing service of the Institute for Genetics, Department Biology I (Ludwig-Maximilians-Universität, München), using BigDye Terminator Ready Reaction Cycles Sequencing Kit v3.1 (Applied Biosystems, Foster City, CA, USA). Sequencing was performed on 6,7 µm DNA probes plus 0.3 µm ITS1F (forward) and 0.3 µm ITS4B (reverse). DNA sequences were aligned pairwise using the BIOEDIT (Bioedit Sequence Alignment Editor for Windows 95/98/NT/XP). Consensus sequences were compared with sequences from the GenBank databsase with BLASTn (National Center for Biotechnology Informations) or UNITE (Kõljalg *et al.*, 2005) and in most case were blasted against both databases.

The anatomotypes are stored in FEA in the TeSAF Department Herbarium of the University of Padua.

3. Results: EM community composition

Morphological, anatomical and molecular investigations revealed a total of 60 anatomotypes. Of these, 8 were assigned to family or ordinal level (*Thelephorales, Boletales, Pezizales, Sebacinaceae, Thelephoraceae*) 19 to genus (*Amphinema* sp., *Boletus* sp., *Cortinarius* sp., *Craterellus* sp., *Hydnum* sp., *Hygrophorus* sp., *Inocybe* sp., *Laccaria* sp., *Lactarius* sp., *Ramaria* sp. *Sebacina sp.*, *Tomentella* sp.), 19 to species [*Byssocorticium atrovirens* (Fr.) Bondartsev & Singer ex Singer, *Cenococcum geophilum* Fr., *Cortinarius bolaris* (Pers.) Fr., *Cortinarius cinnabarinus*

Fr., *Cortinarius infractus* Berk., *Cortinarius inochlorus* Maire, *Genea hispidula* Berk. ex Tul. & C. Tul., *Hygrophorus penarius* Fr. (see Paper I – chapter 3), *Lactarius acris* (Bolton) Gray, *Lactarius pallidus* Pers., *Lactarius rubrocinctus* Fr., *Lactarius subdulcis* Bull., *Lactarius vellereus* (Fr.) Fr., *Piloderma croceum* J. Erikss & Hjortstam, *Ramaria aurea* (Schaeff.) Quél., *Russula illota* Romagn., *Russula mairei* Singer, *Tricholoma acerbum* (Bull.) Vent. *and Tricholoma sciodes* (Pers.) C. Martìn] and *11* [*Fagirhiza arachnoidea* (Brand 1991), *Fagirhiza byssoporoides (description in this Thesis)*, *Fagirhiza cystidiophora* (Brand 1991), *Fagirhiza entolomoides* (description in this Thesis), *Fagirhiza fusca* (Brand 1991), *Fagirhiza lanata* (Brand, 1991), *Fagirhiza oleifera* (Brand 1991), *Fagirhiza pallida* (Brand 1991), *Fagirhiza spinulosa* (Brand 1991), *Fagirhiza stellata* (description in this Thesis), *Fagirhiza vermiculiformis* (Jakucs 1998)] not identified ectomycorrhizae but previously described in details, while 3 remained non-classified (tab. 1). The investigations on ecological features of the EM species are briefly described in the Short Communication (chapter 4).

Fungal taxa, codex in the Herbarium and accession number in the GenBank	Best match sequence	Size bp (pair)	E value	Similarity	Accession number	Source(a)
Amphinema sp. (EDM50)	-	-	-	-	-	-
Boletaceae (EDM51) EU444544	*Boletus aestivalis*	**	3E-73	90%	UDB000941	UNITE
Boletus sp. (EDM13) EU44539	*Boletus rhodoxanthus*	661	0.0	99%	UDB001116	UNITE
Byssocorticium atrovirens (EDM17)	-	-	-	-	-	-
Cenococcum geophilum (EDM1)	-	-	-	-	-	-
Cortinarius ionochlorus (EDM27) EU444542	*Cortinarius inochlorus*	601	0.0	100%	UDB002105	UNITE
Cortinarius (EDM57)sp.	-	-	-	-	-	-
Cortinarius bolaris (EDM12)	-	-	-	-	-	-
Cortinarius cinnabarinus (EDM5)	-	-	-	-	-	-
Cortinarius infractus (EDM62) EU444553	*Cortinarius infractus*	541	0.0	100%	UDB001161	UNITE
Cortinarius sp. (EDM72) EU444551	uncultured ectomycorrhiza (*Cortinarius*)	481	0.0	100%	AY299227	BLAST
Craterellus sp. (EDM41)	-	-	-	-	-	-
EDM47	-	-	-	-	-	-
EDM65	-	-	-	-	-	-
EDM68	-	-	-	-	-	-
Entoloma sp. (EDM36)	-	-	-	-	-	-
Entolomataceae (EDM8)* EU444549	*Entoloma* sp.	901	e-168	91%	UDB000937	UNITE
Fagirhiza arachnoidea (EDM61)	-	-	-	-	-	-
Fagirhiza byssoporoides (EDM55)* EU444550	*Byssoporia terrestris* fruitbody (SR1101 in M)	541	-	99%	-	-
Fagirhiza cystidiophora (EDM33)	-	-	-	-	-	-
Fagirhiza fusca (EDM40)	-	-	-	-	-	-

Fagirhiza lanata (EDM29)	-	-	-	-	-
Fagirhiza oleifera (EDM2)	-	-	-	-	-
Fagirhiza pallida (EDM25)	-	-	-	-	-
Fagirhiza setifera (EDM7)	-	-	-	-	-
Fagirhiza spinulosa (EDM3)	-	-	-	-	-
Fagirhiza stellata (EDM 21)* EU444548	Tomentella subtestacea	661	0.0	92%	UDB000034	UNITE
Fagirhiza vermiculiformis (EDM42)	-	-	-	-	-
Genea hyspidula (EDM32)	-	-	-	-	-
Hydnum sp. (EDM37)	-	-	-	-	-
Hygrophorus sp. (EDM26)	Hygrophorus	**	8e-75	96%	UDB000556	UNITE
Hygrophorus penarius (EDM60)* EU444536	Hygrophorus penarius	481	0.0	100%	UDB000097	UNITE
Inocybe sp. (EDM71) EU444552	Inocybe fuscomarginata	**	0.16	100%	UDB002156	UNITE
Inocybe sp. (EDM22)	-	-	-	-	-
Laccaria sp. (EDM23)	-	-	-	-	-
Lactarius acris (EDM56)	-	-	-	-	-
Lactarius pallidus (EDM6)	-	-	-	-	-
Lactarius rubrocinctus (EDM53)	-	-	-	-	-
Lactarius sp. (EDM48)	-	-	-	-	-
Lactarius subdulcis (EDM4)	-	-	-	-	-
Lactarius vellereus (EDM45)	-	-	-	-	-
Pezicales (EDM 67) EU444547	Peziza sp.	**	3e-57	91%	UDB001572	UNITE
Piloderma croceum (EDM14)	-	-	-	-	-
Ramaria aurea (EDM43)	-	-	-	-	-
Ramaria sp.(EDM58)	-	-	-	-	-
Ramaria sp. (EDM10) EU444537	Albatrellus critstatus	601	1e-91	100%	UDB001761	UNITE

Russula illota (EDM28)	-	-	-	-	-	-
Russula mairei (EDM31)	-	-	-	-	-	-
Sebacina sp.(EDM34) EU444543	Uncultured ectomycorrhiza (*Sebacinaceae*)	541	0.0	95%	AJ879661	BLAST
Sebacinaceae (EDM11) EU444538	*Sebacina epigaea*	541	0.0	94%	UDB000975	UNITE
Thelephoraceae (EDM63)	-	-	-	-	-	-
Thelephoraceae (EDM66)	-	-	-	-	-	-
Thelephorales (EDM64) EU444546	*Tomentellopsis echinospora*	541	0.0	94%	UDB000191	UNITE
Thelephorales (EDM59) EU444545	-	**	-	-	-	-
Tomentella sp. (EDM18) EU444540	*Tomentella cinerascens*	481	0.0	99%	UDB000232	UNITE
Tomentella sp. (EDM19) EU444541	*Tomentella pilosa*	601	0.0	97%	UDB000241	UNITE
Tomentella sp.(EDM46)	-	-	-	-	-	-
Tomentella sp.(EDM70)	-	-	-	-	-	-
Tricholoma acerbum (EDM24)	-	-	-	-	-	-
Tricholoma sciodes (EDM39)	-	-	-	-	-	-

Tab. 1: Ectomycorrhizal anatomotypes and their anatomical and morphological identification. (a) Additional reference are available on the NCBI (www.ncbi.nih.gov/BLAST) or UNITE (www.unite.ut.ee) websites [* Descriptions; ** Partial sequence only].

Fungal taxa	Exploration types	Hydrophobicity
Amphinema sp. (EDM50)	MD fr	hydrophobic
Boletaceae (EDM51)	LD	hydrophilic
Boletus sp. (EDM13)	LD	hydrophobic
Byssocorticium atrovirens (EDM17)	SD	hydrophobic
Cenococcum geophilum (EDM1)	SD	hydrophilic
Cortinarius ionochlorus (EDM27)	MD fr	hydrophobic
Cortinarius (EDM57) sp.	MD fr	hydrophobic
Cortinarius (EDM72) sp.	MD fr	hydrophobic
Cortinarius bolaris (EDM12)	MD fr	hydrophobic
Cortinarius cinnabarinus (EDM5)	MD fr	hydrophobic
Cortinarius infractus (EDM62)	MD fr	hydrophobic
Craterellus sp. (EDM41)	C/SD	hydrophilic
EDM47	SD	hydrophilic
EDM65	MD fr	hydrophobic
EDM68	SD	hydrophobic
Entoloma sp. (EDM36)	MD sm	hydrophobic
Fagirhiza entolomoides (EDM8)*	MD sm	hydrophilic
Fagirhiza arachnoidea (EDM61)	SD	hydrophobic
Fagirhiza byssoporoides (EDM55)*	MD sm	hydrophobic
Fagirhiza cystidiophora (EDM33)	SD	hydrophilic
Fagirhiza fusca (EDM40)	SD	hydrophilic
Fagirhiza lanata (EDM29)	MD sm	hydrophilic
Fagirhiza oleifera (EDM2)	C/SD	hydrophilic
Fagirhiza pallida (EDM25)	SD	hydrophilic
Fagirhiza setifera (EDM12)	SD	hydrophilic
Fagirhiza spinulosa (EDM3)	SD	hydrophilic
Fagirhiza stellata (EDM21)*	MD sm	hydrophobic
Fagirhiza vermiculiformis (EDM42)	MD sm	hydrophilic
Genea hyspidula (EDM32)	SD	hydrophilic
Hydnum sp. (EDM37)	MD fr	hydrophobic
Hygrophorus sp. (EDM26)	C	hydrophilic
Hygrophorus penarius (EDM60)*	SD	hydrophilic
Inocybe sp. (EDM71)	MD mat	hydrophobic
Inocybe sp. (EDM22)	SD	hydrophilic
Laccaria sp. (EDM23)	MD sm	hydrophilic
Lactarius acris (EDM56)	MD sm	hydrophilic
Lactarius pallidus (EDM6)	MD sm	hydrophilic
Lactarius rubrocinctus (EDM53)	MD sm	hydrophilic
Lactarius sp. (EDM48)	C	hydrophilic
Lactarius subdulcis (EDM4)	MD sm	hydrophilic
Lactarius vellereus (EDM45)	MD sm	hydrophilic
Pezizales (EDM 67)	SD	hydrophilic
Piloderma croceum EDM14)	MD fr	hydrophobic
Ramaria aurea (EDM43)	MD mat	hydrophobic

Ramaria sp.(EDM58)	MD mat	hydrophobic
Ramaria sp. (EDM10)	MD mat	hydrophobic
Russula illota (EDM28)	C	hydrophobic
Russula mairei (EDM31)	C	hydrophilic
Sebacina sp.(EDM34)	SD	hydrophilic
Sebacinaceae (EDM11)	SD	hydrophilic
Thelephoraceae (EDM63)	MD sm	hydrophobic
Thelephoraceae (EDM66)	MD sm	hydrophobic
Thelephorales (EDM64)	MD fr	hydrophobic
Thelephorales (EDM59)	MD fr	hydrophobic
Tomentella sp. (EDM18)	MD fr	hydrophobic
Tomentella sp. (EDM19)	MD sm	hydrophilic
Tomentella sp.(EDM46)	MD sm	hydrophilic
Tomentella sp.(EDM70)	SD	hydrophilic
Tricholoma acerbum (EDM24)	MD fr	hydrophobic
Tricholoma sciodes (EDM39)	MD fr	hydrophobic

Tab. 2: Exploration types of the anatomotypes and the relationship with the hydrophobicity [C= contact type - SD= short distance; MD sm= medium distance smooth; MD fr= medium distance fringe; MD mat= medium distance mat; LD= long distance. * Descriptions].

4. Short characterization of the species not described up to now on *Fagus sylvatica* L. (according to Agerer 1991)

Here it is reported a short characterization of the species not described up to now on *Fagus sylvatica* L. To identify the most important morphological and anatomical features, it is used the synoptic key in www.deemy.de by Agerer & Rambold 1998, 2004-2007.

Amphinema sp. (EDM50), Figs. 1-4

Basidiomycota, Basidiomycetes, Polyporales, Atheliaceae

Colour: reddish orange, brownish-red, older part ochre. - Ramification (Fig. 1): irregularly-pinnate, dichotomous-like. - Shape: sinuous - Mycorrhizal surface: woolly. - Rhizomorphs (Fig. 4): orange-reddish, ramified repeatedly into smaller filaments, type A (undifferentiated with hyphae rather loosely woven and of uniform diameter). - Emanating hyphae (Fig. 3): present but not specifically distributed, with open anastomoses, with a short bridge or bridge almost lacking (contact-clamp), sometimes granulate and with thicker wall, intrahyphal hyphae with clamps present, too;

membranaceously yellowish. - Outer mantle (Fig. 2): plectenchymatous type C (gelatinous matrix between the hyphae, membranaceously yellowish). - Middle and inner mantle: plectenchymatous with matrix, membranaceously yellowish. - Exploration type: medium distance subtype fringe. - Hydrophobic attitude.

It can be supposed that this anatomotype belongs to the genus *Amphinema* due to the typical anastomoses and the thick-walled hyphae already reported in the descriptions of *A. byssoides*.

Boletaceae (EDM51), Figs. 8, 9, 75-76

Basidiomycota, Agaricomycetes, Agaricomycetidae, Boletales, Boletaceae

Colour: whitish pink. - Ramification: monopodial-pinnate, monopodial-pyramidal. - Shape: straight; Mycorrhizal surface (Fig. 8): smooth. - Rhizomorphs (Fig. 76): whitish, with smooth margin, type F (highly differentiated - thick hyphae forming mostly a core, septa often partially or completely dissolved), with thick matrix. - Outer mantle (Fig. 75): plectenchymatous type C (gelatinous matrix between the hyphae). - Middle and inner mantle (Fig. 9): plectenchymatous with thick matrix. - Exploration type: long distance. - Hydrophilic attitude.

It can be supposed that this anatomotype belongs to this family, for the typical differentiated rhizomorph. This species showed also a similarity of 90% with *Boletus aestivalis* (Paulet) Fr. (sequences in Unite, see Tab.1).

Boletus sp. (EDM13), Figs. 10, 77-79

Basidiomycota, Agaricomycetes, Agaricomycetidae, Boletales, Boletaceae

Colour: brownish with whitish-granulate surface, when older blackish and whitish-granulate surface more widespread. - Ramification: monopodial-pinnate, monopodial-pyramidal. - Shape: straight, not inflated cylindric. - Mycorrhizal surface (Fig. 10): glistening. - Rhizomorphs (Fig. 77): the same colour of the mantle, infrequent; type F (differentiated; thick hyphae forming a core, septa complete or sometimes enlarged, type E), with nodia and conical young-side branches present, crystals on the peripheral hyphae. - Emanating hyphae: generally present, granulate hyphae (cystidia-like endcells, with small crystals on the surface) membranaceously brownish. - Outer mantle (Fig. 78): plectenchymatous type A/B (ring-like arrangement of hyphal bundles or hyphae rather irregularly arranged and no special pattern discernible), with crystals on the mantle layer. – Middle mantle layer (Fig. 80): pseudoparenchymatous epidermoid type. - Inner mantle (Fig. 79):

plectenchymatous ring-like. - Matrix present in each mantle layer and also in the rhizomorphs. - Exploration type: long distance. - Hydrophobic attitude.

It can be supposed that this anatomotype belongs to this family, for the typical differentiated rhizomorph. The molecular investigations attested that this species showed a similarity of 99% with *Boletus rodoxanthus* (Krombh.) Kallenb (sequences in Unite, see Tab. 1).

***Cortinarius ionochlorus* (EDM27), Figs. 11, 12, 81- 85**

Basidiomycota, Agaricomycetes, Agaricomycetidae, Agaricales, Cortinariaceae

Colour: brownish green with yellowish hydrophobic surface not uniformly distributed; when older brownish and only the rhizomorphs yellow. - Ramification: irregularly-pinnate, dichotomous-like. - Shape: straight, not inflated, cylindric - Mycorrhizal surface (Fig. 11): silvery. - Rhizomorphs (Fig. 85): yellowish, frequent, without specific origin, type A (undifferentiated with hyphae rather loosely woven and uniform diameter). - Outer mantle (Fig. 81): plectenchymatous type A/B (ring-like arrangement of hyphal bundles or hyphae rather irregularly arranged and no special pattern discernible). – Middle mantle layers (Figs. 82, 84) plectenchymatous ring-like. - Inner mantle (Fig. 83): plectenchymatous, ring-like. Membranaceously brownish. - Sclerotia (Fig.12): infrequent, green-yellowish, elongated-irregular, on the mantle and laterally on the rhizomorphs. - Exploration type: medium distance subtype fringe. - Hydrophobic attitude.

Montecchio *et al.* (2001) described this species on *Quercus ilex* L. The structure of the mantle layers and of the rhizomorph are similar to the specimen here reported. The molecular analyses confirmed a similarity of 100% with *Cortinarius ionochlorus* R. Maire (UDB002105 in Unite, see Tab. 1).

***Cortinarius* sp. (EDM57), Figs. 13, 14, 86, 87**

Basidiomycota, Agaricomycetes, Agaricomycetidae, Agaricales, Cortinariaceae

Colour: whitish brown. - Ramification: irregularly-pinnate, dichotomous-like. - Shape: sinuous, bent, not inflated, cylindric. - Mycorrhizal surface (Fig. 13): silvery and densely stringy. - Rhizomorphs (Fig. 87): whitish, type A (undifferentiated with hyphae rather loosely woven and of uniform diameter), frequent, origin not specific. - Outer mantle (Fig. 14): plectenchymatous type A/B (ring-like arrangement of hyphal bundles or hyphae rather irregularly arranged and no special pattern discernible). - Middle mantle (Fig. 86): plectenchymatous. - Inner mantle:

plectenchymatous/pseudoparenchymatous. - Exploration type: medium distance fringe subtype. - Hydrophobic attitude.

It can be supposed that this anatomotype belongs to this genus, in particular for the habitus and for the typical undifferentiated rhizomorph. Also the mantle structure can be related to this genus.

***Cortinarius infractus* (EDM62), Figs. 15, 88, 89**

Basidiomycota, Agaricomycetes, Agaricomycetidae, Agaricales, Cortinariaceae

Colour: whitish. - Ramification: irregularly- pinnate, dichotomous-like. - Shape: sinuous, bent, not inflated, cylindric.- Mycorrhizal surface: very woolly (Fig. 15). – Rhizomorphs: whitish, type A (undifferentiated with hyphae rather loosely woven and of uniform diameter, with crystals and matrix), very frequent, origin not specific. - Outer mantle (Fig. 88) plectenchymatous type C (gelatinous matrix between the hyphae, with crystals). - Middle mantle (Fig. 89): plectenchymatous with matrix. - Inner mantle: plectenchymatous with matrix. - Exploration type: medium distance fringe subtype. – Hydrophobic attitude.

It can be supposed that this anatomotype belongs to this genus, in particular for the habitus, for the typical undifferentiated rhizomorph. Also the mantle structure can be related to this genus. The species was classified thanks to molecular tools. The analyses showed the best similarity (of 100%) with *Cortinarius infractus* Berk. (UDB001161 in Unite, see Tab. 1).

***Cortinarius* sp. (EDM72), Figs. 16, 91, 92**

Basidiomycota, Agaricomycetes, Agaricomycetidae, Agaricales, Cortinariaceae

Colour: yellowish orange. - Ramification: irregularly-pinnate, dichotomous-like. - Shape: sinuous, bent, not inflated, cylindric. - Mycorrhizal surface: very cottony (Fig. 16). - Emanating hyphae (Fig. 91): very frequent anastomoses, open with a short bridge or bridge almost lacking, with clamps; rhizomorphs not observed. - Outer mantle (Fig. 90): plectenchymatous type A (ring-like arrangement of hyphal bundles). - Middle mantle: plectenchymatous. - Inner mantle (Fig. 92): plectenchymatous. - Exploration type: medium distance subtype fringe. – Hydrophobic attitude.

It can be supposed that this anatomotype belongs to this family, in particular for the habitus. Also the mantle structure can be related to this genus and the emanating hyphae (Agerer 2006). The molecular analyses showed a similarity of 100% with a species that belongs to the same genus and collected in a similar ecosystem (uncultered ectomycorrhiza *Cortinarius* sp. AY299227 in

GenBank, see Tab. 1).

Craterellus sp. (EDM41), Figs. 93, 94

Basidiomycota, Agaricomycetes, Cantharellales, Cantharellaceae

Colour: white. - Ramification: monopodial-pinnate. - Shape: not inflated, cylindric. – Mycorrhizal surface: smooth and opaque. - Outer mantle (Fig. 93): pseudoparenchymatous type L (angular cells) with rare oily droplets. - Middle and inner mantle (Fig. 94): pseudoparenchymatous. – Exploration type: contact type to short distance. - Hydrophilic attitude.

It can be supposed that this anatomotype belongs to this genus, because of the absence of particular anatomical features, for the mantle organisation and for the presence of oily droplets also reported in *Craterellus tubaeformis* (Bull.) Quél. + *Pinus sylvestris* (Mleczko 2004).

EDM47, Figs. 17, 18, 95-98

Colour: brownish, with clear very tips. - Ramification: monopodial-pinnate. - Shape: straight, not inflated, cylindric. - Mycorrhizal surface: very spiny (Fig. 17). - Outer mantle (Figs. 18, 95, 96): pseudoparenchymatous type L/D (angular cells with prominent cystidia type A). – Middle (Figs. 97, 98): mantle and inner mantle: pseudoparenchymatous. - Exploration type: short distance. – Hydrophilic attitude.

Probably this species belongs to the *Ascomycota* and in particular to the genus *Tuber*, due to the presence of the same type of cystidia revealed in our specimen and to the lack of clamps. Further investigations are necessary to classify this ectomycorrhiza.

EDM65, Figs. 19, 99, 100

Colour: whitish and slightly pinkish. - Ramification: monopodial-pinnate. - Shape: straight, not inflated, cylindric. - Mycorrhizal surface: silvery (Fig. 19). - Outer mantle (Fig. 99): plectenchymatous type A/B (ring-like arrangement of hyphal bundles or hyphae rather irregularly arranged and no special pattern discernible). - Middle mantle: plectenchymatous. - Inner mantle (Fig. 100): plectenchymatous. - Exploration type: medium distance subtype fringe.

Further investigations are necessary to classify this ectomycorrhiza.

EDM68, Figs. 20, 21, 101

Colour: whitish and slightly orange. - Ramification: monopodial-pinnate. - Shape: straight not inflated cylindric. - Mycorrhizal surface: silvery (Fig. 20). - Rhizomorphs: whitish, type A (undifferentiated with hyphae rather loosely woven and of uniform diameter, with clamps), very frequent, origin not specific. - Outer mantle (Fig. 101) plectenchymatous type A/B (ring-like arrangement of hyphal bundles or hyphae rather irregularly arranged and no special pattern discernible). - Middle mantle: plectenchymatous. - Inner mantle (Fig. 21): plectenchymatous. - Slight matrix present in the mantle layers. – Exploration type: short distance. – Hydrophobic attitude.

Further investigations are necessary to classify this ectomycorrhiza.

Entoloma **sp. (EDM36), Figs. 65-67**

Basidiomycota, Agaricomycetes, Agaricomycetidae, Agaricales, Entolomataceae

Colour: pinkish withish. - Ramification: irregularly-pinnate, dichotomous-like. - Shape: sinuous. - Mycorrhizal surface (Fig. 65): loosely stringy or loosely wolly. – Rhizomorphs: whitish, type A /B (according to Agerer 1991; 1995; Agerer 1985-2006; Agerer & Rambold 2004-2007; Agerer & Iosifidu 2004). - Emanating hyphae not frequent. - Outer mantle (Fig. 66): plectenchymatous type A (ring-like arrangement of hyphal bundles or hyphae rather irregularly arranged and no special pattern discernible). - Middle and inner mantle plectenchymatous (Fig. 67). - Exploration type: medium distance smooth subtype. – Hydrophobic attitude.

It can be supposed that this anatomotype belongs to the genus *Entoloma*, for the habitus, for the outer mantle structure anf for the uniform-compact, stiff and projecting rhizomorphs, very similar to that reported in other species belongs to the same genus, for example in *Entoloma sinuatum* (Bull. Fr.) Kummer + *Salix* spec. (Agerer 1997; 1998).

Hydnum **sp. (EDM37), Figs. 22-24**

Basidiomycota, Agaricomycetes, Cantharellales, Hydnaceae

Colour: yellowish, with orange very tip. - Ramification: monopodial-pinnate, monopodial-pyramidal. - Shape: straight , not inflated, cylindric. - Mycorrhizal surface: woolly, sometimes covered with soil particles (Fig. 22). - Rhizomorphs: concolourous to mantle, repeatedly ramified into smaller filaments or infrequently at restricted points, smooth or hairy or densely enveloped by hyphae, undifferentiated, hyphae rather losely woven and of uniform diameter (type A) or slightly differentiated, central hyphae somewhat enlarged (type C). - Outer mantle (Fig. 23) plectenchymatous type A (ring-like arrangement of hyphal bundles) with slight matrix and oily droplets. - Middle mantle: plectenchymatous. - Inner mantle (Fig. 24): plectenchymatous/pseudoparenchymatous. - Exploration type: medium distance fringe subtype. - Hydrophobic attitude.

The mantle organisation and the rhizomorph are very similar to the anatomical features reported in *Hydnum rufescens + Picea* (Agerer *et al.* 1996; Kraigher & Agerer 1996).

Hygrophorus sp. (EDM26), Figs. 25, 102, 103

Basidiomycota, Agaricomycetes, Agaricomycetidae, Agaricales, Hygrophoraceae

Colour: brownish chestnut. - Ramification: simple, rarely monopodial-pyramidal. - Shape: straight or bent, not inflated, cylindric. - Mycorrhizal surface (Fig. 25) : smooth sometimes with soil particles. - Outer mantle (Fig. 102) plectenchymatous, type C (gelatinous matrix between the hyphae) clamps very rare. - Middle mantle: plectenchymatous with gelatinous matrix. - Inner mantle (Fig. 103): plectenchymatous. - Exploration type: contact type. – Hydrophilic attitude.

It can be supposed that this species belongs to this genus, for the thick gelatinous matrix, the mantle structure and for the very rare clamps. The anatomical features are similar to that reported in the description of *Hygrophorus penarius* (see Chapter 3), but in contrast to this species the measures of the hyphae are smaller. The molecular analyses confirmed only the genus (see Tab. 1)

Inocybe sp. (EDM 71), Figs. 26, 104-107

Basidiomycota, Agaricomycetes, Agaricomycetidae, Agaricales, Inocybaceae

Colour: brownish-greenish. - Ramification: irregularly-pinnate, dichotomous-like. - Shape: straight or bent, or beaded - Mycorrhizal surface (Fig. 26): woolly. - Rhizomorphs (Fig. 107): concolourous to mantle; sometimes membranaceouly brownish, type A (undifferentiated with hyphae rather loosely woven and of uniform diameter). - Emanating hyphae: with clamps, membranaceously brownish. - Outer mantle (Fig. 104): plectenchymatous type E (hyphae arranged net-like, repeatedly and squarrosely branched) membranaceously brownish. - Middle mantle (Fig. 105): plectenchymatous /pseudoparenchymatous membranaceously brownish. - Inner mantle (Fig. 106):

plectenchymatous /pseudoparenchymatous membranaceously brownish. - The colour of the mantles is darker at patches. - Exploration type: medium distance mat subtype. - Hydrophobic attitude.

It can be supposed that this antomotype belongs probably to this genus for the typical outer mantle and of the rhizomorphs (similar to other species of the same genus reported in www.deemy.de). The molecular analyses confirmed only partially the best similarity with *Inocybe fuscomarginata* Kühner (UDB002156 in Unite, see Tab. 1).

Inocybe **sp. (EDM 22), Figs. 27-29**

Basidiomycota, Agaricomycetes, Agaricomycetidae, Agaricales, Inocybaceae

Colour: brownish grey, with rosy very tip. - Ramification: monopodial-pinnate; monopodial-pyramidal. - Shape: straight, cylindric, not inflated. - Mycorrhizal surface (Fig. 27): loosely cottony. - Emanating hyphae: with clamps. - Outer mantle (Fig. 28): plectenchymatous type E (hyphae arranged net-like, repeatedly and squarrosely branched) membranaceouly brownish . - Middle mantle: plectenchymatous/pseudoparenchymatous, membranaceously brownish. - Inner mantle (Fig. 29): plectenchymatous membranaceouly brownish. - Slightly gelatinous matrix present in the mantle layers. - Exploration type: short distance. - Hydrophilic attitude.

It can be supposed that this antomotype belongs probably to this genus for the outer mantle, similar to that in *Inocybe avellana* + *Shorea* (Ingleby K 1999).

Laccaria **sp. (EDM23), Figs. 30-33**

Basidiomycota, Agaricomycetes, Agaricomycetidae, Agaricales, Hydnangiaceae

Colour: pinkish brown, brownish orange. - Ramification: monopodial-pyramidal. - Shape: straight, or bent, slightly tortuous. - Mycorrhizal surface (Fig. 30): loosely cottony. - Outer mantle (Fig. 31): plectenchymatous type B (hyphae rather irregularly arranged and no special pattern discernible). - Middle mantle (Fig. 32): plectenchymatous whitout pattern. - Inner mantle (Fig. 33): plectenchymatous, ring-like. - Exploration type: medium distance smooth subtype. – Hydrophlic attitude.

It can be supposed that this species belongs to the genus *Laccaria* in particularly for the structure of the mantle layers: the plectenchymatous outer mantle with rather short, obtuse, even finger-like branches and the ring-like arrangement of the hyphae in the inner mantle.

Lactarius sp. (EDM48), Figs. 34-36

Basidiomycota, Agaricomycetes, Russulales, Russulaceae

Colour: pinkish brown. - Ramification: monopodial-pyramidal. - Shape: straight. - Mycorrhizal surface (Fig. 34): smooth. - Outer mantle (Fig. 35): pseudoparenchymatous type Q (epidermoid cells bearing a hyphal net). - Middle mantle: pseudoparenchymatous. - Inner mantle (Fig. 36): pseudoparenchymatous. - Laticifers: present in the mantle, straight and even. – Exploration type: contact type. – Hydrophilic attitude.

This anatomotypes belongs to the genus *Lactarius*, for the presence of laticifers.

Pezizales (EDM67), Figs. 37, 108-111

Ascomycota, Pezizomycetes, Pezizomycetidae, Pezizales, Pezizaceae

Colour: orange-ochre. - Ramification: monopodial-pinnate. - Shape: straight, cylindric or tapering. - Mycorrhizal surface (Fig. 37): smooth, or loosely cottony. - Outer mantle (Figs. 108, 109): type Q, pseudoparenchymatous (epidermoid cells bearing a hyphal net), with prominent cystidia. - Middle mantle: pseudoparenchymatous. - Inner mantle (Fig. 110): pseudoparenchymatous. - Cystidia (Fig. 111): type N (capitate) on the outer mantle. - Cell walls of the angular cells and the net on the outer mantle: thick. – Exploration type: short distance. – Hydrophilic attitude.

There are no ectomycorrhizal species descriptions up to now about members of this family (De Roman *et al.*, 2005). It can be supposed that this anatomotypes belongs to the *Ascomycota* for the typical anatomical features, but the genus is confirmed only partially (see Tab. 1).

Ramaria sp. (EDM10), Figs. 38, 39, 112-114

Basidiomycota, Agaricomycetes, Phallomycetidae, Gomphales, Gomphaceae

Colour: whitish, brownish, yellowish. - Ramification: irregularly-pinnate, dichotomous-like. - Shape: straight, cylindric not inflated, or sinuous. - Mycorrhizal surface (Fig. 38): silvery, fan-like. - Rhizomorphs (Figs. 112, 114): margin not smooth, dividing repeatedly into smaller filaments, slightly differentiated, central hyphae somewhat enlarged (type C). - Outer mantle (Fig. 113): type A/B (ring-like arrangement of hyphal bundles or hyphae rather irregularly arranged and no special pattern discernible) - Middle mantle: plectenchymatous - Inner mantle: plectenchymatous. - Cystidia (Fig. 39): type P as acanthocystidia on the rhizomorphs. - Slight gelatinous matrix and crystals on the rhizomorphs. - Exploration type: medium distance subtype mat. - Hydrophobic attitude.

It can be supposed that this anatomotypes belongs to this genus, for the habitus, the mantle structure with matrix but in particular for the rhizomorphs organisation and for the presence of the typical cystidia (Agerer 2006).

Ramaria sp. (EDM58), Figs. 40, 115-117

Basidiomycota, Agaricomycetes, Phallomycetidae, Gomphales, Gomphaceae

Colour: whitish, pinkish. - Ramification: monopodial-pinnate. - Shape: straight, cylindric not inflated. - Mycorrhizal surface (Fig. 40): hairy, fan-like. – Rhizomorphs: sligthly differentiated, central hyphae somewhat enalarged (type C). - Outer mantle (Fig. 115): type A/B (ring-like arrangement of hyphal bundles or hyphae rather irregularly arranged and no special pattern discernible). - Middle mantle: plectenchymatous. - Inner mantle (Fig. 116): plectenchymatous. - Cystidia not observed. - Sclerotia (Figs. 40, 117): abundant and globular, whitish, with several crystals, on the emanating hyphae. - Exploration type: medium distance subtype mat. - Hydrophobic attitude.

It can be supposed that this anatomotypes belongs to this genus, for the habitus, the mantle structure with matrix but in particular for the rhizomorphs structure (ramaroid), with ampullate hyphae (Agerer 2006).

Sebacina sp. (EDM34), Figs. 41, 42, 118-120

Basidiomycota, Agaricomycetes, Sebacinales, Sebacinaceae

Colour: whitish, pinkish, yellowish. - Ramification: simple. - Shape: straight, cylindric not inflated. - Mycorrhizal surface (Fig. 41): loosely cottony. - Emanating hyphae (Fig. 120): with simple septa, thick cell walls; irregularly inflated or even beaded. - Outer mantle: type E (hyphae arranged net-like, repeatedly and squarrosely branched (Figs. 42; 118); surface of mantle characterized by emanating hyphae, thick wall with rare simple septa. - Middle mantle: plectenchymatous. - Inner mantle (Fig. 119): plectenchymatous. - Cystidia not observed. – Exploration type: short distance. – Hydrophilic attitude.

It can be supposed that this anatomotypes belongs to this genus for morphological features very similar to that fo other specimens and for the anatomy: the hyphae arrangement in the outer mantle with rather shor, obtuse, even finger-like branches is present also in *Sebacina incrustans* (Pers.) Tul. & C. Tul. + *Picea abies*, and for the clampless and smooth typical eamanating hyphae (Agerer 2006).
The molecular analyses confirmed only partially the family (see Tab. 1).

Sebacinaceae (EDM11), Figs. 43-47

Basidiomycota, Agaricomycetes, Sebacinales, Sebacinaceae

Colour: orange. - Ramification: monopodial-pyramidal. - Shape: bent and tortuous, cylindric, not inflated. - Mycorrhizal surface (Fig. 43): very loosely cottony. - Emanating hyphae (Figs. 44, 47): with simple septa, straight or tortuous (thick-walled and straight or ramified, or thin-walled and tortuous). - Outer mantle (Fig. 45): type B, surface of the mantle characterized by emanating hyphae, thicker with rare simple septa - Middle mantle: plectenchymatous. - Inner mantle (Fig. 46): plectenchymatous. - Cystidia (Fig. 47): slightly tapering, often rather similar to ends of normal hyphae, but mostly originating from a pseudoparenchyma. - Sclerotia (Fig. 44): yellowish, on the mantle, infrequent and globular. - Exploration type: short distance. – Hydrophilic attitude.
It can be supposed that this anatomotype belongs to this genus, for the tipical emanating hyphae clampless and smooth and for the presence of the cystidia (Agerer 2006). The molecular analyses confirmed only partially the genus, showing a similarity of 94% with *Sebacina epigaea* (Berk. & Broome) Bourdot & Galzin (UDB000975 in Unite, see Tab. 1).

Thelephoraceae (EDM63), Figs. 48, 121-125

Basidiomycota, Agaricomycetes, Thelephorales

Colour: greenish, brownish, older part blackish, yellowish points. - Ramification: monopodial-pyramidal. - Shape: straight, cylindric not inflated. - Mycorrhizal surface (Fig. 48): densely grainy. - Rhizomorphs (Fig. 125): infrequent, thelephoroid type, with repeatedly ramified, densely entwined and glued peripheral hyphae, with nodia, conical structures, clamps. - Outer mantle (Fig. 122): pseudoparenchymatous star-like, with prominent cystidia, slightly gelatinous matrix on the star surface and small crystals. - Middle mantle (Fig. 123): pseudoparenchymatous. - Inner mantle (Fig. 124): pseudoparenchymatous. - Cystidia (Fig. 121): bent or curved with thick walls, sickle-shaped. -

Mantle layers and emanating hyphae membranaceously dark brow. - Exploration type: medium distance smooth subtype. – Hydrophobic attitude.

It can be supposed that this anatomotypes is a member of the *Thelephorales*, for the star-like structure in the outer mantle, with prominent cystidia, for the typical thelephoroid rhizomorphs and for the organs colour (Agerer 2006).

Thelephoraceae **(EDM66), Figs. 49, 51, 52, 126**

Basidiomycota, Agaricomycetes, Thelephorales

Colour: dark brown, black. - Ramification: monopodial-pyramidal. - Shape: straight, cylindric not inflated. - Mycorrhizal surface (Fig. 49): densely grainy. - Rhizomorphs (Fig. 53): infrequent, thelephoroid type, with irregularly sinuous peripheral hyphae, nodia and conical structures, clamps. - Outer mantle (figs. 51): angular cells, bearing mounds of roundish cells (type K), angular cells with thick cell walls. - Middle mantle (Fig. 126): pseudoparenchymatous rosette-like. - Inner mantle (Fig. 52): plectenchymatous; mantle layers and emanating hyphae membranaceously brownish. - Exploration type: medium distance subtype smooth. – Hydrophobic attitude.

It can be supposed that this anatomotype is a member of the *Thelephorales*, for the outer mantle, for the typical thelephoroid rhizomorphs and for the habitus (the colour and the ramification; Agerer 2006).

Thelephorales **(EDM64), Figs. 5, 68-71**

Basidiomycota, Agaricomycetes, Agaricomycetidae, Thelephorales

Colour: orange-ochre, or yellow-brown with red spots. - Ramification: monopodial-pinnate, monopodial-pyramidal. - Shape: bent, sinuous. - Mycorrhizal surface (Fig. 5): loosely cottony. - Rhizomorphs (Fig. 71): brownish-orange, ramified repeatedly into smaller filaments type A (undifferentiated with hyphae rather loosely woven and of uniform diameter), anastomoses closed by a simple septum, with a short bridge or bridge almost lacking (contact-septum), with membranaceously brownish to yellowish content; emanating hyphae not frequent, with clamps. - Outer mantle (Fig. 68): plectenchymatous type A/B (ring-like arrangement of hyphal bundles or hyphae rather irregularly arranged and no special pattern discernible). - Middle (Fig. 69) and inner mantle (Fig. 70) plectenchymatous. - Exploration type: medium distance subtype fringe. - Hydrophobic attitude.

It can be supposed that this anatomotype belongs to this order and probably to the genus *Tomentellopsis (Thelephorales)* for the hyphae arrangement of the outer mantle and for the rhizomorph structure. This species showed also a similarity of 94% with the species *Tometellopsis echinospora* (Ellis) Hjortstam (sequence in Unite, see Tab. 1).

Thelephorales (EDM59), Figs. 6, 7, 72-74

Basidiomycota, Agaricomycetes, Agaricomycetidae, Thelephorales

Colour: pink, or pinkish-white; Ramification: irregularly-pinnate, dichotomous-like. - Shape: sinuous. - Mycorrhizal surface (Fig. 6): loosely cottony. - Rhizomorphs: orange, with ramification repeatedly into small filaments or, frequently at restricted points, type A (undifferentiated with hyphae rather loosely woven and uniform diameter), (Fig. 73) anastomoses closed by a simple septum, with a short bridge or bridge almost lacking (contact-septum), membranaceously red-pinkish (but not uniform, Fig. 7). - Emanating hyphae not frequent. - Outer mantle (Fig. 72): plectenchymatous type A/B (ring-like arrangement of hyphal bundles or hyphae rather irregularly arranged and no special pattern discernible). - Middle and inner mantle (Fig. 74): plectenchymatous. - Exploration type: medium distance subtype fringe. - Hydrophobic attitude.

It can be supposed that this anatomotype belongs to this order and probably to the genus *Tomentellopsis (Thelephorlaes)* for the hyphae arrangement of the outer mantle and for the rhizomorph structure (Agerer 2006).

Tomentella sp. (EDM18), Figs. 54, 55, 127-129

Basidiomycota, Agaricomycetes, Thelephorales, Thelephoraceae

Colour: orange-reddish, brownish. - Ramification: monopodial-pyramidal. - Shape: straight, cylindric not inflated - Mycorrhizal surface (Fig. 54): loosely woolly. - Emanating hyphae: without clamps. - Rhizomorphs (Fig. 55) frequent, type B (undifferentiated; margins rather smooth; hyphae compactly arranged and of uniform diameter), anastomoses open with a long bridge. - Outer mantle (Figs. 127, 128): type B (hyphae rather irregularly arranged and no special pattern discernible), with the surface covered by emanating hyphae without clamps. - Middle mantle: plectenchymatous. - Inner mantle (Fig. 129): plectenchymatous; mantle layers and emanating hyphae membranaceously brownish. - Exploration type: medium distance fringe subtype. – Hydrophobic attitude.

It can be supposed that this anatomotype is a member of the *Thelephoraceae*, for the outer mantle and for the emanating hyphae clampless and for the habit (Agerer 2006), and for the undifferentiated rhizomorphs not frequent for this genus, but present in *Tomentella galzinii* Bourd + *Quercus*, too (www.deemy.de). Furthermore the anatomical features are very similar to that identified as *Tomentella* sp. (EDM70).

The molecular analyses confirmed the family, showing a similarity of 99% with *Tomentella cinerascens* (P. Karst.) Höhn. & Litsch. (UDB000241 in Unite, see Tab. 1)

Tomentella **sp. (EDM19), Figs. 56-58**

Basidiomycota, Agaricomycetes, Thelephorales, Thelephoraceae

Colour: brown. - Ramification: monopodial-pyramidal. - Shape: straight, cylindric, not inflated or tapering. - Mycorrhizal surface (Fig. 56): loosely grainy and stringy. - Rhizomorphs: infrequent, thelephoroid type, with irregularly sinuous peripheral hyphae, nodia and conical structures, whith clamps. - Outer mantle: pseudoparenchymatous, angular cells bearing prominent clavate cystidia. - Middle mantle (Fig. 58): between plectenchymatous and pseudoparenchymatous. - Inner mantle: plechtenchymatous with clamps. – Cystidia (Fig. 57): clavate, connected with a clamp to the mantle. - Mantle layers and emanating elements membranaceously brownish. – Exploration type: medium distance smooth subtype. – Hydrophilic attitude.

It can be supposed that this anatomotypes is a member of the *Thelephorales*, for the outer mantle, similar to that reported in the group of *Tomentella subtestacea* and *Tomentella pilosa* (Agerer 2006), and also for the typical thelephoroid rhizomorphs and for the cystidia. The molecular analyses showed a similarity of 97% with *Tomentella pilosa* (Burt) Bourdot & Galzin (UDB000241 in Unite, see Tab. 1).

Tomentella **sp. (EDM46), Figs. 59, 61**

Basidiomycota, Agaricomycetes, Thelephorales, Thelephoraceae

Colour: brown, blackish with whitish very tip. - Ramification: monopodial-pyramidal. - Shape: straight, cylindric, not inflated. - Mycorrhizal surface (Fig. 59): loosely cottony. - Rhizomorphs: infrequent, thelephoroid type, with irregularly sinuous peripheral hyphae, nodia and conical structures, with clamps. - Outer mantle (Fig. 61): plectenchymatous, type B (hyphae rather irregularly arranged and no special pattern discernible). - Middle mantle: plectencymatous. - Inner mantle: plechtenchymatous. - Mantle layers and emanating elements membranaceously brownish. –

Exploration type: medium distance smooth subtype. – Hydrophilic attitude.

It can be supposed that this anatomotype is a member of the *Thelephorales*, for the outer mantle very similar to that in *Tomentella ferruginea* (Pers.) Pat. + *Fagus sylvatica* (Raidl & Müller 1996; Raidl 1998), for the similar thelephoroid rhizomorphs reported in this description.

Tomentella sp. (EDM70), Figs. 60, 62-64, 130

Basidiomycota, Agaricomycetes, Incertae sedis, Thelephorales, Thelephoraceae

Colour: brownish, blackish - Ramification: simple, monopodial-pyramidal. - Shape: straight, cylindric not inflated. - Mycorrhizal surface (Fig. 60): loosely cottony. - Emanating hyphae (Fig. 62): without clamps, sometimes epimembranaceously brownish. - Rhizomorphs (Fig. 63): frequent, type B (undifferentiated; margins rather smooth; hyphae compactly arranged and of uniform diameter), with protuberances on the hyphae. - Outer mantle (Fig. 64): type B (hyphae rather irregularly arranged and no special pattern discernible), surface covered by emanating hyphae without clamps - Middle mantle: plectenchymatous - Inner mantle (Fig. 130): plectenchymatous. - Mantle layers and emanating hyphae membranaceously brownish. - Exploration type: short distance. – Hydrophilic attitude.

This anatomotypes showed morphological and anatomical similarity with *Tomentella* sp. (EDM 18). For this reason it can be supposed that both belong to the same genus.

References

AGERER R(1986). Studies on ectomycorrhizae II. Introducing remarks on characterization and identification. *Mycotaxon* 26: 473–492.

AGERER R (1991) Characterization of ectomycorrhizae. In: NORRIS JR, READ DJ, VARMA AK, (eds) Techniques for the Study of Mycorrhiza. UK Academic Press, London, pp 25-73.

AGERER R (1987-2006). Colour atlas of ectomycorrhizae. 1th -13th del. Einhorn, Schwäbisch Gmünd, Germany.

AGERER R (1994). Characterization of ectomycorrhiza. In Techniques for Mycorrhizal Research, (J. R. Norris, D. Read & A. K. Varma (eds): 25–73. Academic Press, London.

AGERER R (1995). Anatomical characteristics of identified ectomycorrhizas: an attempt towards a natural classification. In Varma K, Hock B (eds) mycorrhiza: structure, function, molecular biology and biotechnology. Springer, Berlin Heidelberg, New York: 685-734.

AGERER R (1999). Anatomical characteristics of identified ectomycorrhizas: an attempt towards a natural classification. In:Mycorrhiza: structure, function, molecular biology and biotechnology (A. Varma & B. Hock, eds): 633–682. 2nd edn. Springer-Verlag, Berlin

AGERER R (2006). Fungal relationship and structural patterns of their ectomycorrhizae. *Mycological Progress* 5(2): 67-107.

AGERER R, IOSIFIDOU P (2004). Rhizomorph structures of Hymenomycetes: a possibility to test DNA-based phylogenetic hypotheses? In Agerer R, Piepenbring M, Blanz P (eds) Frontiers in Basidiomycote Mycology, IHW-Verlag, Eching: 249-302.

AGERER R, RAMBOLD G (1998). DEEMY, a DELTA-based system for characterization and Determination of EctoMYcorrhizae. Version 1.1.

AGERER R, RAMBOLD G (2004-2007). [first posted on 2004-06-01; most recent update: 2007-03-##]. DEEMY – An Information System for Characterization and Determination of Ectomycorrhizae. www.deemy.de – München, Germany.

AGERER R (2001). Exploration types of ectomycorrhizal mycelial systems. A proposal to classify mycorrhizal mycelial systems with respect to their ecologically important contact area with the substrate. *Mycorrhiza* 11: 107-114.

AGERER R, KRAIGHER H, JAVORNIK B (1996). Identification of ectomycorrhizae of *Hydnum rufescens* on Norway spruce and the variability of the ITS region of *Hydnum rufescens* and H. repandun (Basidiomycetes). Nova Hedwigia 63: 183-194.

BRAND F (1991). Ektomycorrhizen an *Fagus sylvatica*. Charaktersierung und Identifizierung, ökologische Kennzeichnung und unsterile Kultivierung. Libri Botanici, IHW-Verlag : 1-229.

CAIRNEY JWG, CHAMBERS SM (eds) (1999). Ectomycorrhizal fungi: key genera in profile. Springer-Verlag, Berlin

DE ROMAN M, CLAVERIA V, DE MIGUEL AM (2005). A revision of the description of ectomycorrhiza published since 1961. *Mycological Research* 109 (10):1063-1104.

DOMINIK T (1969). Key to ectotrophic mycorrhizae. Folia Forestalia Polonica 15: 309–328.

GARDES M AND BRUNS TD (1993). ITS primers with enhanced specificity for basidiomycetes – application to the identification of mycorrhizae and rusts. *Molecular Ecology* 2: 113–118.

GOODMAN DM, DURALL DM, TROFYMOW JA, BERCH SM (EDS) (1996–2000). A Manual of concise Descriptions of North American Ectomycorrhizae. *Mycologue Publications*, Sidney, BC.

HAUG I, WEBER R, OBERWINKLER F, TSCHEN J (1994). The mycorrhizal status of Taiwanese trees and the description of some ectomycorrhizal types. *Trees* 8: 237–253.

INGLEBY K, MASON PA, LAST FT, FLEMING LV (1990). Identification of Ectomycorrhizas. Institute of Terrestrial EcologyResearch Publication No. 5. HMSO, London.

INGLEBY K (1999). *Inocybe avellana* Horak + *Shorea leprosula* Miq. *Descr Ectomyc* 4: 55-60.

KÕLJALG U, LARSSON K-H, ABARENKOV K, NILSSON RH, ALEXANDER IJ, EBERHARDT U, ERLAND S, HOILAND K, KJOLLER R, LARSSON E, PENNANEN T, SEN R, TAYLOR AFS, TEDERSOO L, VRALSTAD T, URSING BM (2005). UNITE: a database providing web-based methods for the molecular identification of ectomycorrhizal fungi. New Phytologist 166: 1063-1068.

KRAIGHER H, AGERER R (1996). Hydnum rufescenc. In Agerer R (ed) Colour Atlas of Ectomycorrhizae, plate 92, Einhorn-Verlag, Schwäbisch Gmünd.

MLECZKO P (2004). *Craterellus tubaeformis* (Bull.) Quél. + *Pinus sylvestris* L. *Descr Ectomyc* 7/8: 29-36.

MOLINA R, MASSICOTTE H, TRAPPE JM (1992). Specificity phenomena in mycorrhizal symbioses: Community-ecological consequences and practical implications. In: Routledge AMF, ed. Mycorrhizal functioning, an integralive plant–fungal process. New York, USA: Chapman & Hall, Inc., 357–423.

MONTECCHIO L, ROSSI S, GRENDENE A, CAUSIN R (2001). *Cortinarius ionochlorus* R. Maire + *Quercus ilex* L. *Descr Ectomyc* 5: 35-40.

RAIDL S (1998). *Tomentella ferruginea. In Agerer R (ed) Colour Atlas of Ectomycorrhizae*, plate 137, Einhorn-Verlag, Schwäbisch. Gmünd.

RAIDL S, MÜLLER WR (1996). *Tomentella ferruginea* (Pers.) Pat. + *Fagus sylvatica* L. *Descr Ectomyc* 1: 161-166.

TAYLOR ANDY FS, ALEXANDER I (2005). The ectomycorrhizal symbiosis: life in the real world. Mycologist, Volume 19, Part 3 August 2005 p. 102-112.

TEDERSOO L, SUVI T, LARSSON E, KÕLJALG U (2006). Diversity and community structure of ectomycorrhizal fungi in a wooded meadow. *Mycological Research* IIO(2006) pp. 734-748.

UNESTAM T (1991). Water repellency, mat formation, and leaf-stimulated growth of some ectomycorrhizal fungi. *Mycorrhiza*, 1: 13-20.

ZAK JC (1973). Classification of Ectomycorrhizae. In Ectomycorrhizae, their Ecology and Physiology (G. C. Marks & T. T. Kozlowski, eds): 43–78. Academic Press, New York.

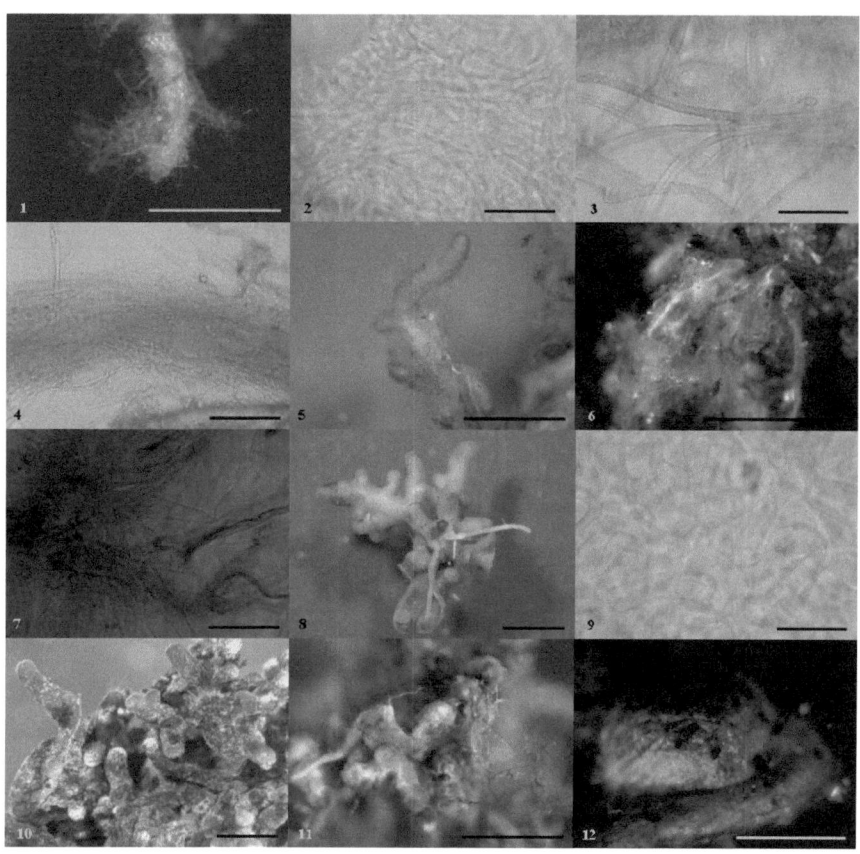

Figs. 1-4: *Amphinema* sp. (EDM50); Fig. 1. Habit, 1 bar =1 mm; Fig. 2: Outer mantle, 1 bar = 20 μm; Fig. 3. Emanating hyphae, 1 bar= 20 μm; Fig. 4. Rhizomorph, 1 bar= 8 μm. Fig. 5: *Thelephorales* (EDM64), Habit, 1 bar = 1 mm. Figs. 6-7: *Thelephorales* (EDM59); Fig. 6. Habit 1 bar = 1 mm; Fig. 7. Red-pinkish content, 1 bar = 20 μm. Figs. 8-9: *Boletaceae* (EDM51); Fig. 8. Habit, 1 bar = 1 mm; Fig. 9. Inner mantle, 1 bar = 20 μm. Fig. 10: *Boletus* sp. (EDM13), Habit, 1 bar = 1mm. Figs. 11-12: *Cortinarius inochlorus* (EDM27); Fig. 11. Habit, 1 bar = 1 mm; Fig. 12. Sclerotia, 1 bar = 0.5 mm.

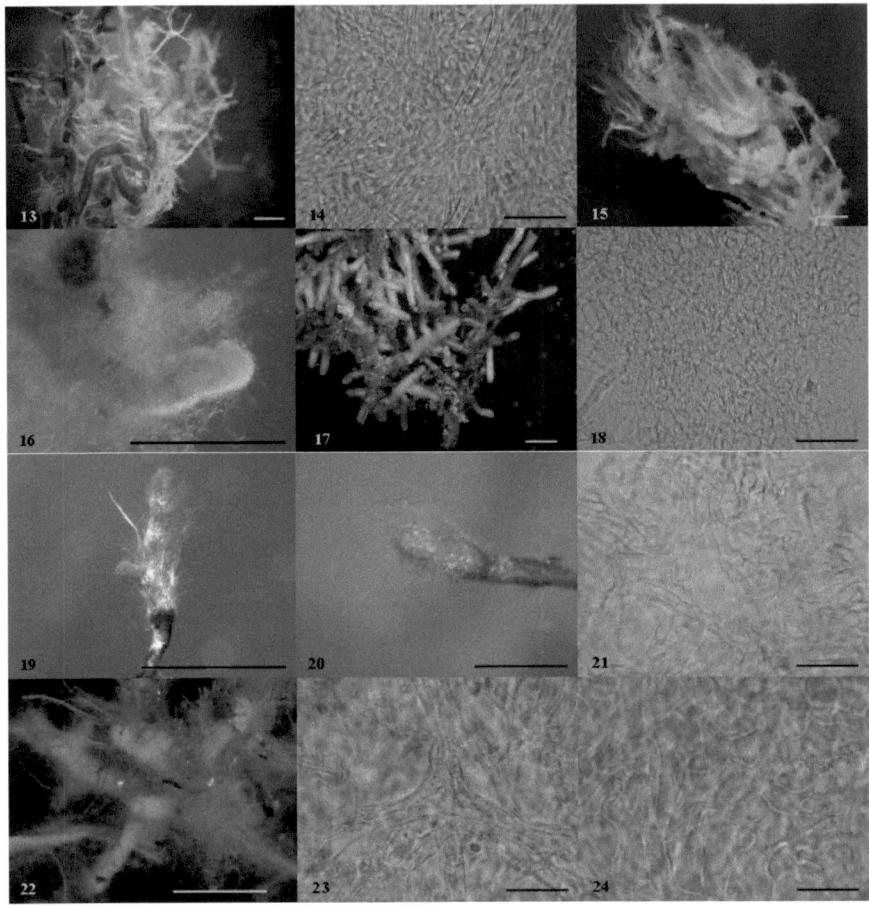

Figs. 13-14: *Cortinarius* sp. (EDM57); Fig. 13. Habit, 1 bar = 1mm; Fig. 14. Outer mantle, 1 bar = 20 µm. Fig. 15: *Cortinarius infractus* (EDM62), Habit, 1 bar = 1mm. Fig. 16: *Cortinarius* sp. (EDM72), Habit, 1 bar = 1mm. Fig. 17-18:Unidentified etcomycorrhiza (EDM47); Fig. 17. Habit, 1 bar = 1 mm; Fig. 18. Outer mantle, 1 bar = 8 µm. Fig: 19: Unidentified ectomycorrhiza (EDM 65), Habit, 1 bar = 1mm. Figs. 20-21: Unidentified ectomycorrhiza (EDM 68); Fig. 20. Habit, 1 bar = 1mm; Fig. 21. Inner mantle, 1 bar = 20 µm. Figs. 22-24: *Hydnum* sp. (EDM37); Fig. 22. Habit, 1 bar = 1mm; Fig. 23. Outer mantle, 1 bar 20 µ; Fig. 24. Inner mantle, 1 bar = 20 µm.

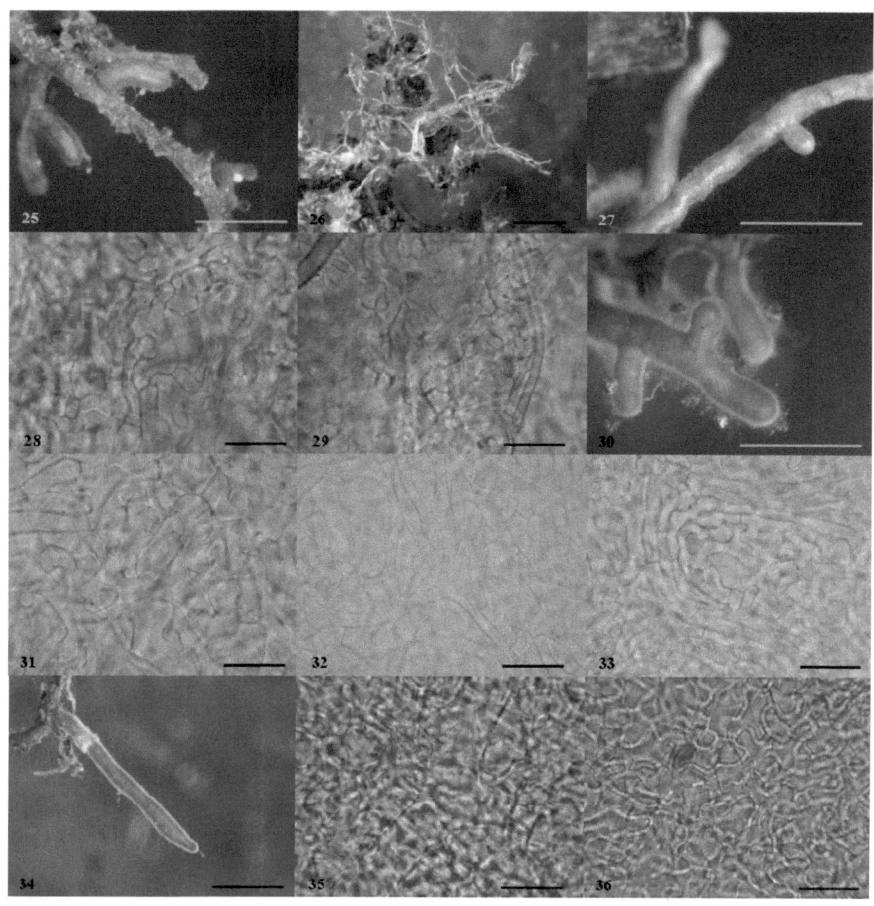

Fig. 25: *Hygrophorus* sp. (EDM26), Habit, 1 bar = 1 mm. Fig. 26: *Inocybe* sp. (EDM71), Habit, 1 bar = 1 mm. Figs. 27-29: *Inocybe* sp. (EDM22); Fig. 27. Habit, 1 bar = 1 mm; Fig. 28. Outer mantle, 1 bar = 20 μm; Fig. 29. Inner mantle, 1 bar = 20 μm. Figs. 30-33: *Laccaria* sp. (EDM23); Fig. 30. Habit, 1 bar = 1mm; Fig. 31. Outer mantle, 1 bar = 20 μm; Fig. 32. Middle mantle, 1 bar = 20 μm; Fig. 33. Inner mantle, 1 bar = 20 μm. Figs. 34-36: *Lactarius* sp. (EDM48); Fig. 34. Habit, 1 bar = 1 mm; Fig. 35. Outer mantle, 1 bar = 20 μm; Fig. 36. Inner mantle, 1 bar 20 μm.

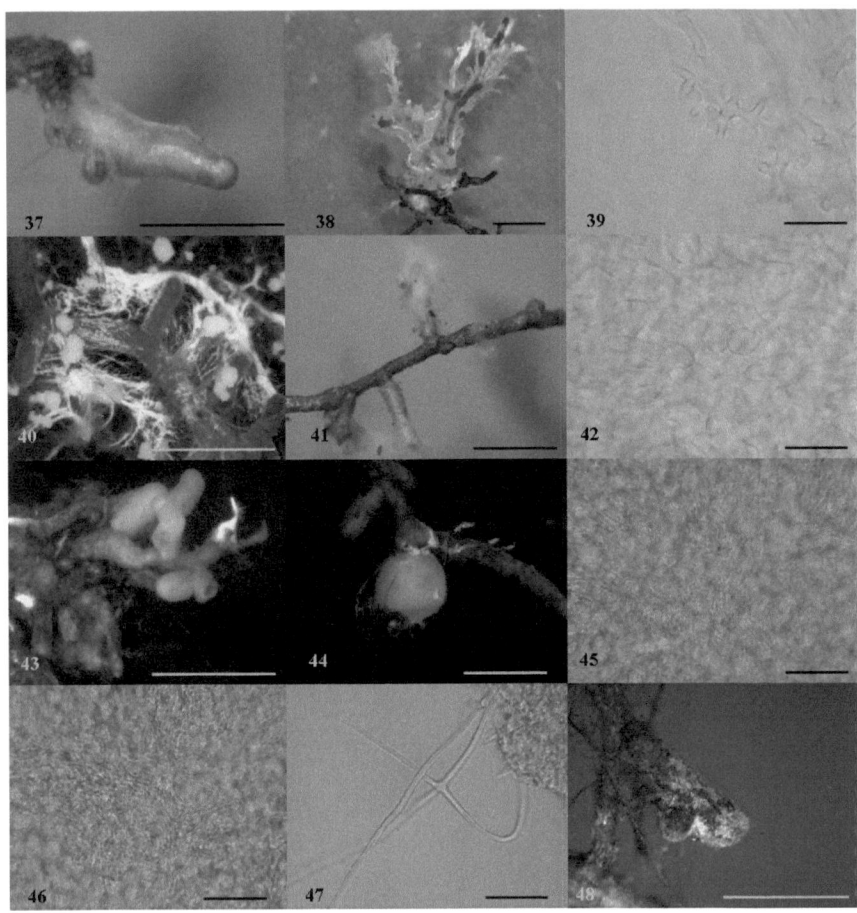

Fig. 37: *Pezizales* (EDM67), Habit, 1 bar = 1 mm. Figs. 38-39: *Ramaria* sp. (EDM10); Fig. 38. Habit, 1 bar = 1 mm; Fig. 39. Cystidia, 1 bar = 20 μm. Fig. 40: *Ramaria* sp. (EDM58), Habit, 1 bar = 1 mm. Figs. 41-42: *Sebacina* sp. (EDM34); Fig. 41. Habit, 1 bar =1 mm; Fig. 42. Outer mantle, 1 bar = 20 μm. Figs. 43-47: *Sebacinaceae* (EDM11); Fig. 43. Habit, 1 bar – 1 mm; Fig. 44. Sclerotia, 1 bar – 1 mm; Fig. 45. Inner mantle, 1 bar – 8 μm; Fig. 46. Middle mantle, 1 bar = 8 mm; Fig. 47. Emanating hyphae and cystidia. Fig. 48: *Thelephoraceae* (EDM63), Habit, 1 bar = 1 mm.

Figs. 49-53: *Thelephoraceae* (EDM66); Fig. 49. Habit, 1 bar = 1 mm; Fig. 50. Outer mantle, 1 bar = 20 μm; Fig. 51. Outer mantle, 1 bar 20 μm; Fig. 52. Middle mantle, 1 bar = 20 μm. Fig. 53. Rhizomorph, 1 bar = 20 μm. Figs. 54-55: *Tomentella* sp. (EDM18); Fig. 54. Habit, 1 bar = 1 mm; Fig. 55. Emanating hyphae, 1 bar = 20 μm. Figs. 56-58: *Tomentella* sp. (EDM19); Fig. 56. Habit, 1 bar = 1 mm; Fig. 57. Cystidia, 1 bar = 20 μm; Fig. 58. Middle mantle, 1 bar = 20 μm. Fig. 59: *Tomentella* sp. (EDM46), Habit, 1 bar = 1 mm. Fig. 60: *Tomentella* sp. (EDM70), Habit, 1 bar = 0.5 mm.

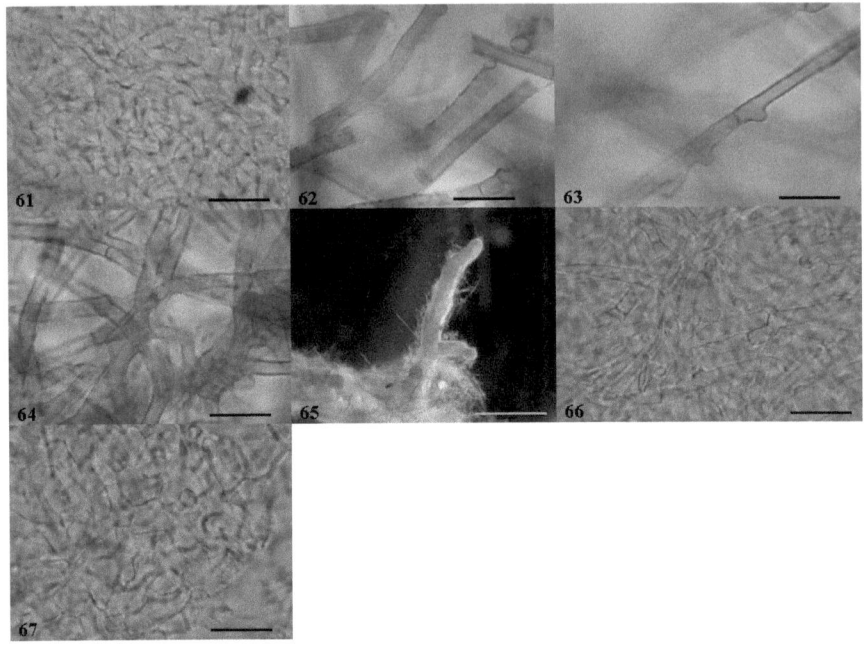

Fig. 61. *Tomentella* sp. (EDM46), Outer mantle, 1 bar = 20 µm. Figs. 62- 64: *Tomentella* sp. (EDM70) emanting hyphae, 1 bar = 20 µ; Fig. 64. Outer mantle, 1 bar = 20 µm. Figs. 65-67: Entoloma sp. (EDM36). Fig. 65. Habit, 1 bar = 1 mm; Fig. 66. Outer mantle, 1 bar = 20 µm; Fig. 67. Inner mantle, 1 bar = 20 µm.

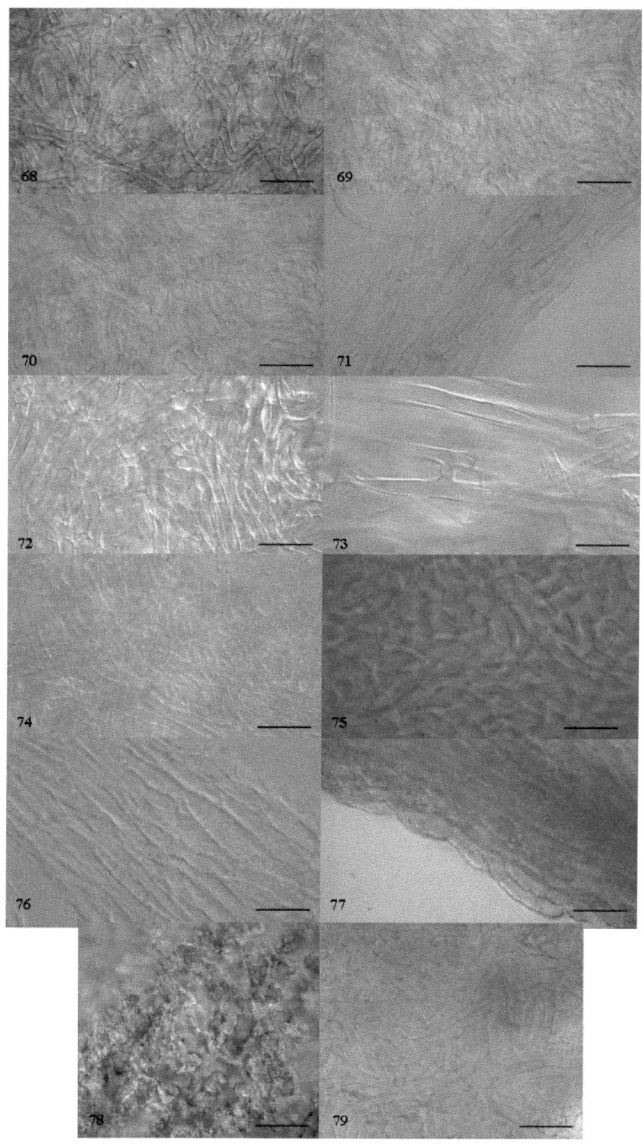

Figs. 68-71: *Thelephorales* (EDM64); Fig. 68. Outer mantle; Fig. 69. Middle mantle; Fig. 70. Inner mantle; Fig. 71. Rhizomorph. Figs. 72-74: *Thelephorales* (EDM59); Fig. 72. Outer mantle; Fig. 73. Anastomoses closed by a simple septum; Fig. 74. Inner mantle. Figs. 75-76: *Boletaceae* (EDM51); Fig. 75. Outer mantle; Fig. 76. Rhizomorph. Figs. 77-79: *Boletus* sp. (EDM13); Fig. 77. Rhizomorph; Fig. 78. Outer mantle; Fig. 79. Inner mantle. Bar for all Figs. = 20 μm.

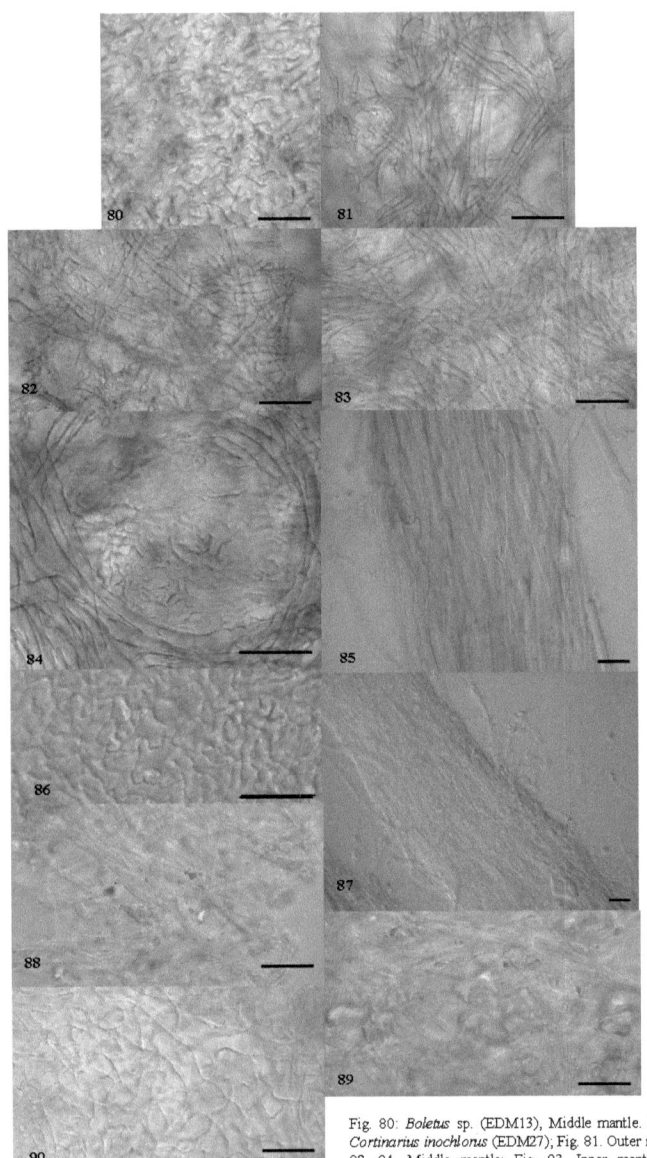

Fig. 80: *Boletus* sp. (EDM13), Middle mantle. Figs. 81-85: *Cortinarius inochlorus* (EDM27); Fig. 81. Outer mantle; Figs. 82, 84. Middle mantle; Fig. 83. Inner mantle; Fig. 85. Rhizomorph. Figs. 86-87: *Cortinarius* sp. (EDM57); Fig. 86. Middle mantle; Fig. 87. Rhizomorph. Figs. 88-89: *Cortinarius infractus* (EDM62); Fig. 88. Outer mantle; Fig. 89. Middle mantle. Fig. 90: *Cortinarius* sp. (EDM72), Outer mantle. Bar for all Figs. = 10 μm

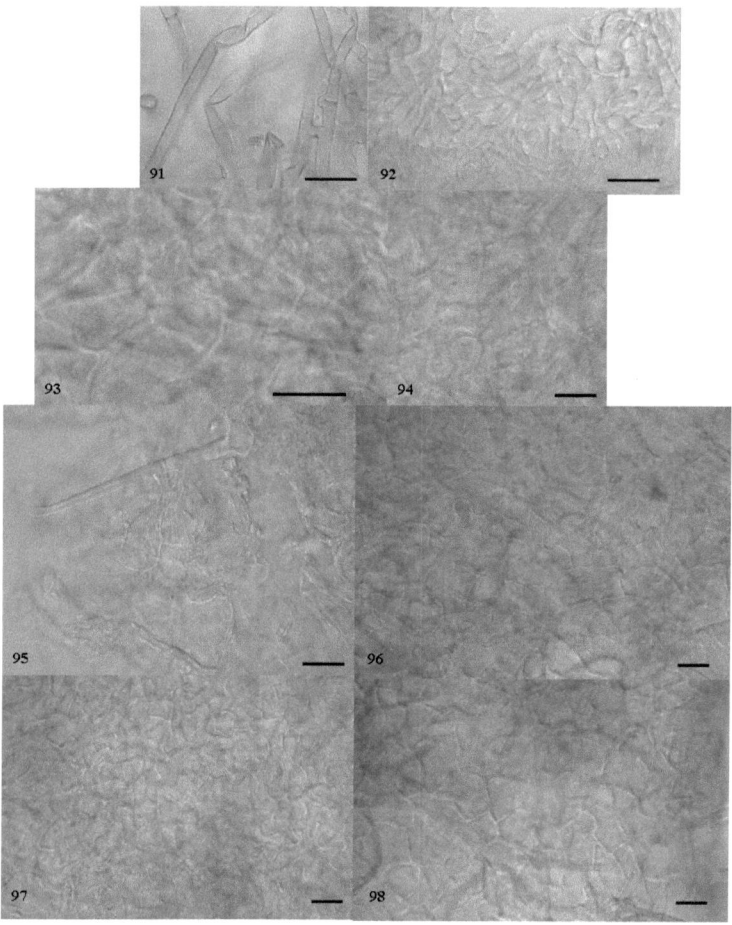

Figs. 91-92: *Cortinarius* sp. (EDM72); Fig. 91. Emanating hyphae; EDM92. Outer mantle. Figs. 93-94: *Craterellus* sp. (EDM41); Fig. 93. Outer manlte; Fig. 94. Middle mantle. Fig. 95-98: Unidentified ectomycorrhiza (EDM47); Fig. 95. Outer mantle and cystidia; Fig. 96. Outer mantle; Fig. 97. Middle mantle; Fig. 98. Inner mantle. Bar for all Figs. = 10 μm.

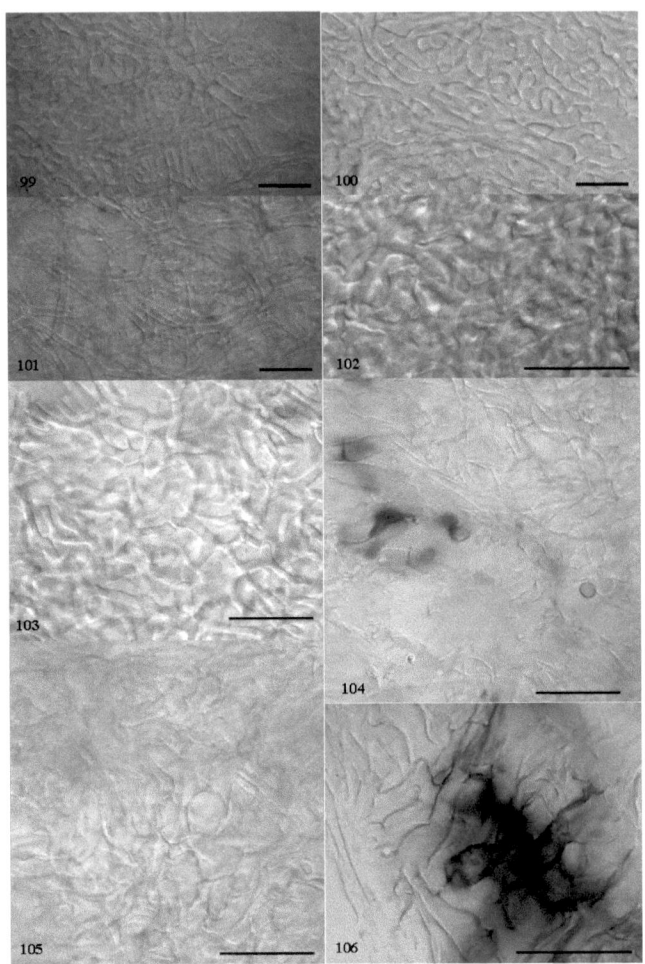

Figs. 99-100: Unidentified ectomycorrhiza (EDM65); Fig. 99. Outer mantle; Fig. 100. Inner mantle. Fig. 101: Unidentified ectomycorrhiza (EDM68), Outer mantle. Figs. 102-103: *Hygrophorus* sp. (EDM26); Fig. 102. Outer mantle; Fig. 103. Inner mantle. Figs. 104-106: *Inocybe* sp. (EDM71); Fig. 104. Outer mantle; Fig. 105. Middle mantle; Fig. 106. Inner mantle. Bar for all Figs. = 20 μm.

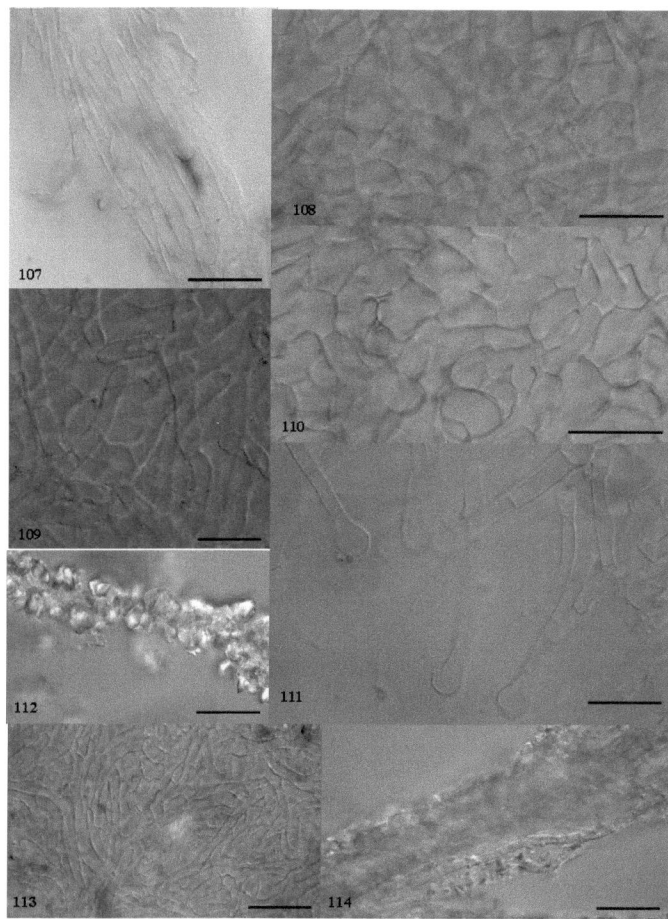

Fig. 107: *Inocybe* sp. (EDM71, Rhizomorph. Figs. 108-111: Unidentified ectomycorrhiza (EDM67), Outer mantle; Fig. 109. Mantle surface; Fig. 110. Inner mantle; Fig. 111. Cystidia. Figs. 112-114: *Ramaria* sp. (EDM10); Fig. 112. Crystals on the hyphae; Fig. 113. Outer mantle; Fig. 114. Rhizomorph. Bar for all Figs. = 10 μm.

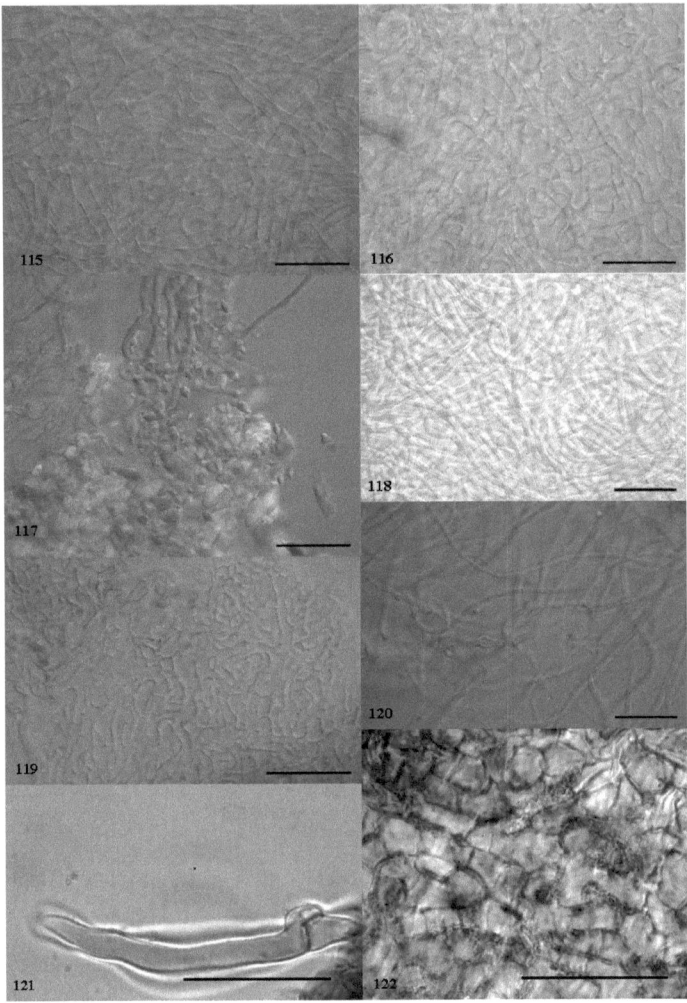

Figs. 115-117: *Ramaria* sp. (EDM58); Fig. 115. Outer mantle; Fig. 116. Inner mantle; Fig. 117: Young sclerotia. Figs. 118-120: *Sebacina* sp. (EDM34); Fig. 118. Mantle surface; Fig. 119. Inner mantle; Fig. 120. Emanating hyphae. Figs. 121-122: *Thelephoraceae* (EDM63); Fig. 121. Cystidia. Fig. 122. Outer mantle. Bar for all Figs. = 10 μm.

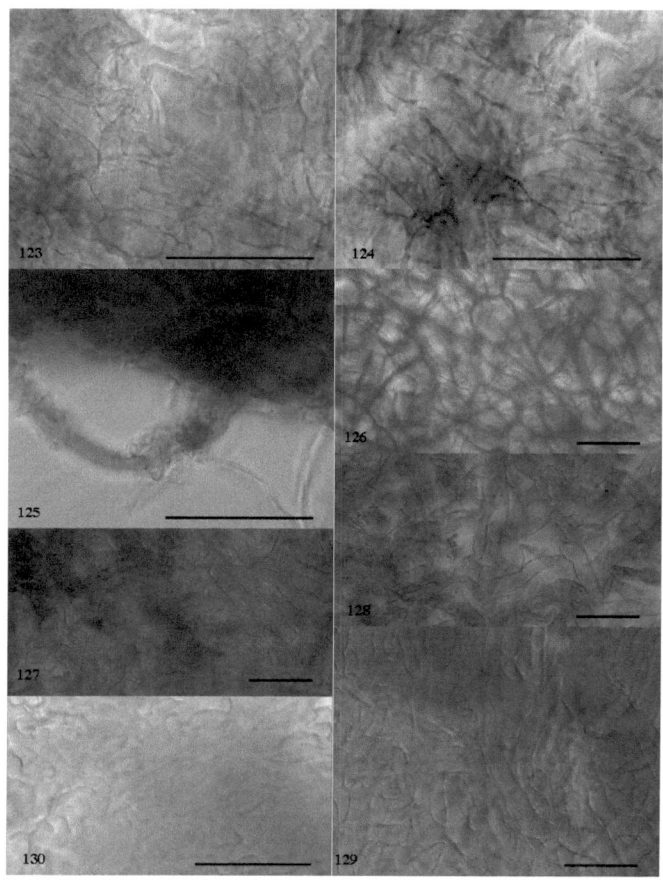

Figs. 123-125: *Telephoraceae* (EDM63); Fig. 123. Middle mantle; Fig. 124. Inner mantle; Fig. 125. Nodia of the rhizomorph. Fig. 126: *Telephoraceae* (EDM66), Middle mantle. Figs. 127-128: *Tomentella* sp. (EDM18), Outer mantle; Fig. 128. Mantle surface. Fig. 130: *Tomentella* sp. (EDM70), Inner mantle. Fig. 129: *Tomentella* sp. (EDM18), Inner mantle. Bar for all Figs. =10 µm.

CHAPTER 3

Hygrophorus penarius on beech: between mutualism and parasitism?

Submitted the 29th Jenuary 2008 to Mycorrhiza

Erika Di Marino[1,2]*, *Lucio Montecchio*[1], *Reinhard Agerer*[3]

[1] Università degli Studi di Padova Agro-Forestry Systems and Land Use (TeSAF) Department, v.le dell'Università, 16; I-35020 Legnaro (PD) – Italy

[2] Centro di Ecologia Alpina - Viote del Monte Bondone; I-38070 Sopramonte (TN), Italy

[3] Organismic Biology: Mycology, Department Biology and GeoBio-Center[LMU], University of München, Menzinger Str., 67; D-80638, Germany

*) Corresponding author
Submitted to Mycorrhiza

Abstract. The mycorrhizae of *Hygrophorus penarius* on *Fagus sylvatica* are described and compared to other species of the same genus and with unidentified mycorrhizae known from literature. The mycorrhiza of *H. penarius* is very similar to that of *Entoloma saepium* on *Rosa* sp., in that the fungus appears to prevent the formation of a root meristem and invades young root cells, in a parasitic-like behaviour, but dissimilar to *E. saepium* that destroys and digests the root apex. The Hartig net is not formed, although a very thick mantle is present that is composed of infrequently clamped hyphae embedded in a gelatinous matrix. To get more information about its behaviour the stable carbon and nitrogen isotope ratios of its mycorrhizae were studied, revealing a negative $\delta^{15}N$ value, similar to that of non-mycorrhizal roots and of many typical ectomycorrhizae. $\delta^{13}C$ values did not reveal important information. Due to the special type of interaction between the fungus and the root, a parasitic-like strategy for the studied *H. penarius* on beech can be hypothesized - at least under the considered growth conditions.

Key words: ectomycorrhiza, mutualism, parasitism, induced resistance, isotopes, ^{15}N, ^{13}C, *Fagus sylvatica*.

1 Introduction

The definition of mycorrhiza is based on the concept of symbiosis *sensu lato* (de Bary, 1887) between a fungus and a plant, and in most cases mycorrhizae consist of a mutualistic relationship, where both partners benefit (Smith & Read 1997). Notwithstanding this, in ectomycorrhizae it is often difficult to discern the exact nutrient pathways between the two symbionts and the direct and indirect benefits derived by each of the symbionts, as both physiological and environmental features dynamically influence the relationship (Ahmadijan &, 1997; Peterson *et al.*, 2004). On this subject, the term "balanced mycorrhizae" well defines the conditions in which both organisms obtain essential resources through reciprocal exchanges (Brundrett 2004), the terms "exploitive mycorrhization" (Brundrett 2004) or "reciprocal exploitations" (Herre *et al.*, 1999) are used when a unbalanced, mainly plant-directed, nutrient flow happens, while "reciprocal parasitism" (Peterson *et al.*, 2004) can be used when the nutrient flow is mainly directed to the fungus, gradually destabilizing the partnership and approaching a parasitic behaviour. Anatomical examples of parasitic-like ectomycorrhizal infections are known on *Fagus sylvatica*, *Rosa* spp. and *Prunus* spp. (Götsche 1972; Agerer & Waller 1993; Kobayashi & Hatano 2001).

From a behavioural point of view it was demonstrated that, in a mutualistic partnership, the balance of antagonism between plant and fungus could gradually tilt in favour of the second when the changed environmental or partner's features (i.e. suboptimal ecological conditions, cell age) allow it to parasitize few or many cells, thanks to its phenotypic plasticity (Kuldau & Yates 2000; Jumpponen 2001; Sieber 2002). This depiction, anyway, is far from generalization, changing not only with the two involved species, but at least with their genotype and age, and with the changing environmental characters (Lu *et al.* 2004; Rodrigues *et al.* 2004). Furthermore, the fungal protrusion into single plant cells, not always defines a parasitic activity towards the whole plant: "induced resistance" is a physiological state of enhanced defensive capacity elicited by specific environmental stimuli (i.e. the exposition to a nonpathogenic organism), whereby the plant's non-specific and innate defences (i.e. phytoalexins production, cell's lignification and/or suberization) strengthen against subsequent pathogenic challenges (Horsfall & Cowling 1980; Bailey 1985; Van Driesche & Bellows 1996; Sticher *et al.*, 1997; van Loon *et al.*, 1998).

To this respect, the characterization of the nutritional attitude of an ectomycorrhizal fungus is of major importance, and the presence, abundance and ratio of stable isotopes [^{15}N (δ^{15}N); ^{13}C (δ^{13}C)] in fungi may be a useful approach to define a mutualistic behaviour (Gebauer & Taylor 1999;

Hobbie *et al.* 1999; 2001; 2002; Kohzu *et al.*, 1999).

The main goals of the research were to comprehensively describe the ectomycorrhiza of *Hygrophorus penarius* on *Fagus sylvatica*, and to compare its carbon and nitrogen isotopic signature to that of known ectomycorrhizal fungi and roots.

2 Material and Methods

The methods for characterization of EM are comprehensively described by Agerer (1991). Fresh material was studied regarding morphology, colour of hyphae, and chemical reactions; material fixed in FEA was applied to produce mantle preparations as well as for longitudinal sections. The drawings were made using a ZEISS Axioskop with Normarski's Interference Contrast, at a magnification of 2000x with the aid of a drawing mirror, transferred on a transparent paper by Indian ink drawing devices, and finally reduced in magnification.

Identification was possible by comparison of nuclear rDNA ITS sequences obtained from the mycorrhizal root tip and from the fruitbody. DNA extraction, PCR and sequencing methods follow Tedersoo *et al.* (2006). GenBank accession number of the EM sequence is EU444536. The reference specimens of the mycorrhizae and of the fruitbodies are deposited in PD (TeSAF Department herbarium; Holmgren *et al.* 1990). The ectomycorrhizal material was collected in Italy, Trento province (Trentino-Alto Adige Region), Val di Non, district Denno (46°14' N; 10° 57' E), beech coppice cut in 2004, mesic condition, 1050 m a.s.l., mineral soil, humus type Dysmull, limestone substrate, soil pH 5.3-6.0, N_{tot} 13,9-21,8, C/N 16-18, moisture 52-69,7, C_{org} 226-395g/Kg. Myc. isol E. Di Marino, 12.06.2006 (older ontogenetical stage EDM 60 in FEA , Agerer 1991, Reference specimen) and 9.05.2007 (younger ontogenetical stage).

For isotopic studies, fruitbodies of the following provenances were used. *Lactarius acris* (Bolton) Gray, Germany, Bavaria, Berchtesgarten National Park 1080 m a.s.l., leg. 12.09.1979 (in M); *Hygrophorus russula* (Schaeff.) Quél., Italy, Ferrara, Bosco Mesola, leg. 08.12.1999 (in PD); *H. penarius Fr.* collection 1, Italy, Pordenone, Barcis, 450 m a.s.l., leg. 08.12.1999 (PD); collection 2, Italy, Reggio Calabria, Ciciarella, leg. 06.10.1999 (in PD); collection 3, Italy, Bologna, Parco dei Gessi, leg 26.10.1994 (in PD).

For analyses of ^{15}N and ^{13}C content, non-mycorrhizal dried roots, mycorrhizal tips and fruitbodies were ground to fine powder and analysed by a combined element analyser (EA3000, Euro Vector instruments and software, Milano, Italy) and isotope ratio mass spectrometer (IsoPrime, GV-Instruments, Manchester, UK) for their C and N concentrations as well as for their ^{13}C and ^{15}N values. All isotope ratios were expressed in δ notation relative to the standards of PeeDee Belemnite (PDB; Dawson *et al.* 2002) for carbon and atmospheric N_2 for nitrogen. The analyses were performed by the Centre of Life and Food Sciences Department of Ecology/Plant Ecophysiology (TMU Freising, Germany).

3 Results

Mycorrhiza of *Hygrophorus penarius* on *Fagus sylvatica* L.

Morphological characters (Figs. 1a-c, 8-11)

Mycorrhizal systems solitary or in few numbers, not or infrequently ramified, then irregulary pinnate, with distinct opaque mantle surface; cortical cells visible through the almost completely transparent mantle, mantle easily removable form the root as a gelatinous cap; mycorrhizal surface smooth, when older covered by sand and soil particles and hyphae; hydrophilic, of the short-distance exploration type (Agerer 2001), transparent and colourless, the root below the mantle whitish and changing to black with age; side branches 0.5-4 mm long and 0.5-0.8 mm diam., straight or bent, not inflated, cylindrical. Emanating hyphae abundant and concentrated proximally in older tips; rhizomorphs and sclerotia lacking.

Anatomical characteristic in plan views (Figs. 2-5, 6a, b, 7, 13-17)

Outer mantle layers (Figs. 2, 3, 6b, 12, 13) with a surface covered by a very thick gelatinous matrix and with many soil particles and crystals; very loosely plectenchymatous, with a well extended gelatinous matrix between the hyphae (mantle type C, according to Agerer 1991, 1995, Agerer 1987-2006, Agerer & Rambold 2004-2007); hyphae cylindrical, sometimes constricted at septa, inflated at clamps, often slightly ampullate at both sides of the septum; angles between hyphal junctions ca. 90°, rarely 45° or 120°, colourless; thinner hyphae 1-2 µm diam. with 0.1 µm wide walls, with infrequent simple septa and clamps; hyphae most frequently 4-7(9) µm diam. and with (10)30-60(100) µm long cells, walls thin, others with (1)2(5) µm wide walls, infrequent open anastomoses - *Middle mantle layers* (Figs. 4, 5) densely plectenchymatous, without pattern, with a very extended gelatinous matrix, hyphal surface smooth, but with ca. 5x10 µm large rhomboid crystals; walls 0.2-0.5 µm, cells (15)30-(60) µm long, (2)4-5 µm diam.; thicker fraction of hyphae more infrequent than in outer layer, with 2-3 µm thick walls, cells (5)7-9 µm diam., clamps present, simple or medallion-like, with rare open anastomoses. *Inner mantle layers* (Fig. 6a, 16, 17) densely plectenchymatous without pattern, colourless with gelatinous and thick matrix, hyphae (2.5)3.5-5(7) µm., with infrequent simple septa and smaller, ca. 3x7 µm large, rhomboid crystals.

Anatomical characters of emanating elements

Rhizomorphs not observed. - *Emanating hyphae* (Figs. 7, 14, 15) tortuous or irregularly inflated or even beaded, angle of ramification acute or approximately 90°, ramification adjacent to septum, hyphal ends simple or tortuous or screw-like; cell walls of tips as thick as remaining walls or thicker, with normal clamps or inflated without or with infrequent simple septa, sometimes out of the mantle but full immersed in the matrix, with crystals and soil particles; hyphae (3)4-6(10) μm in diam., those lacking a gelatinous matrix reveal (0.5)2(3.5) μm thick walls, others have 2-3.5 μm wide walls and a matrix of 10-12(15) μm width, some lack a cell lumen and consist exclusively of cell wall material; in lateral view clamps with a hole, more or less than a semicircle, constricted at contact point to subtending hyphal cell, hyphae at septa even; anastomoses open with a short bridge or bridge lacking, or closed by a simple septum and almost without a hyphal bridge; *cystidia* and *chlamydospores* not observed.

Anatomical characters, longitudinal section
Mantle plectenchymatous with gelatinous matrix, very wide, compact, (30)70(100) μm thick, different layers not discernable, colourless. *Epidermal cells* reacting against the hyphae (Fig. 21), by the formation of thicker cell walls (Figs. 19, 20) and sometimes by condensed tannins within the cells, forming apparently a "zone of defence" against the frequent intracellularly growing hyphae (Figs. 18, 19); outer root cell layers seem to be necrotic; root meristem and Hartig net not visible, vascular tissue reaching almost the tip of the mycorrhiza.
In younger ontogenetical stages mantle with 25-30(40) μm less wide in comparison to that of the older stage, meristem not totally digested, but intracellular hyphae present in the cells of the meristem (Fig. 23).

Colour reaction in different reagents
Mantle preparations: cotton blue: n.r. (no reaction); ethanol 70%: n.r.; FEA: n.r.; iron (II)sulphate: slightly greyish; lactic acid: n.r.; Melzer's reagent: n.r.; Congo red: n.r.; sulpho-vanillin: n.r., or with reddish spots within the matrix; KOH 10%: n.r.; guaiac: n.r..

Autofluorescence
Whole mycorrhiza: UV 254 nm: lacking; UV 366 nm: violet-blue. *Mantle in section:* UV-filter 340-380 nm: slightly bluish; blue filter 450-490 nm: slightly yellowish; filter 530-560 nm: slightly reddish.

Isostope ratios ($\delta^{13}C$ and $\delta^{15}N$)
The results obtained comparing *Hygrophorus penarius* and *Lactarius acris* (Bolton) Gray fruitbodies of different provenances and ectomycorrhizae collected in the same time and in the same

site are reported in tab. 1, showing higher δ^{15}N values for *L. acris*, with *H. penarius* values close to the ones arising from non mycorrhizal root tips. Other trials did not give appreciable differences between species.

4 Discussion

The main anatomical features of *Hygrophorus penarius* on *Fagus sylvatica* are, firstly, the hydrophilic, colourless, transparent mantle with a gelatinous matrix containing variously thick hyphae with irregularly thickened walls forming a loosely plectenchymatous structure, secondly, intracellular hyphae that grow into the cells of the root tip, apparently preventing the root from forming a meristem, with the consequence that the elements of the central stele reach almost the very tip of the root, and, thirdly, the lack of a Hartig net with root cells reacting by thickening of their walls and producing tannic substances.

The mycorrhizae of the genus *Hygrophorus* that have been characterized to date show some similarities with those of *H. penarius*, at least with respect to some mantle features.

The mantle of *Hygrophorus lucorum* Kalchbr. + *Larix decidua* is with 60-70 µm also very wide (Treu 1990), but at least the outer mantle layers are pseudoparenchymatous, not gelatinous and emanating hyphae are not covered by a matrix. The mantle hyphae of the species described here are completely immersed in a gelatinous matrix, as previously observed for EM of *Hygrophorus olivaceoalbus* (Fr.) Fr. (sub nomine *Piceirhiza gelatinosa*, Agerer 1987-2006; Berg *et al.*, 1989; Gronbach & Agerer 1986) + *P. abies*, but in contrast to *H. penarius*, the mantle hyphae of *H. olivaceoalbus* EM are arranged labyrinthine-like (Gronbach & Agerer 1986; Berg *et al.*,1989). Intracellular hyphae can also occur in *H. olivaceoalbus*, but they originate from the Hartig net that is often composed of several hyphal rows (Gronbach & Agerer 1986; Haug & Pritsch 1992). Whether hyphae are growing within the very tip of this EM is unknown, as longitudinal sections through the meristem are not studied yet (Gronbach & Agerer 1986; Haug & Pritsch 1992). The outer mantle of *H. pustulatus* (Gronbach 1988; 1989) is not gelatinous and forms rhizomorphs. Gronbach (1988) reports for this species intracellular hyphae that can possess clamps within cortex cells, but they are restricted to cells enveloped by the Hartig net and are not growing within the very tip of the root.

Additional EM species are known to form, mainly in old stages, intracellular structures (haustoria), but otherwise with features typical for EM associations, as reported for *Elaphomyces granulatus* Fr. (*Piceirhiza glutinosa,* Gronbach 1988, 1989; Haug & Pritsch 1992), *Piceirhiza ascosphinctrina* (Haug & Pritsch 1992), *P. glutinosaesimilis* (Berg 1989) and *P. guttata* (Berg 1989; Gronbach 1988, 1989). Haustoria may be a lifelong attribute of several species with a typical Hartig net and an intact root meristem, as in *Lactarius acris* (Brand 1991), *Quercirhiza ectendotrophica* (Azul *et al.*, 2001),

Russula mairei Sing. (Brand 1991) and *Tricholoma scioides* (Secr.) Mart (Brand 1991). Stronger anatomical modifications were demonstrated in *Entoloma saepium* (Noulet & Dass.) Richon & Roze on both *Rosa* spp. and *Prunus* spp. (Waller & Agerer 1993), and in *Entoloma clypeatum* f. *hybridum* on *Rosa multiflora* (Kobayashi & Hatano 2001). Here, the outer cell layers and the Hartig net appeared modified, digested and incomplete, the distal parts of the tip degenerated, including the meristems and most of the cortical cells, and the intracellular hyphal presence left only cell remnants behind, indicating a parasitic behaviour. Götsche (1972) observed comparable features studying an unclassified EM with an ectendotrophic stage on *F. sylvatica*, where the fungus digested progressively the epidermal and the meristemic cells, the mantle lost its consistency and suberin layers were formed. However, detailed anatomical studies were not provided.

Hygrophorus penarius on *F. sylvatica* appears poised between mutualism and parasitism. It lacks the Hartig net typical of ectomycorrhizal mutualists. Its intracellular colonization of plant cells (Fig. 22), the partial digestion of cortical cells, and the prevention of meristematic tissue at least in later ontogenetical stages are also more typical of parasites than of mutualists. The plant reaction to this colonization consisted of cell wall thickenings and tannic substances filling the cells.

To evaluate the mutualistic activity of *H. penarius*, its $\delta^{15}N$ and $\delta^{13}C$ isotope signatures were measured in root tips, mycorrhizae and fruitbodies. We compared these data with isotopic signatures in the mutualistic species *Lactarius acris*.

The *H. penarius* EM showed a negative $\delta^{15}N$ value (-4‰), quite different from the mean value of 5,7‰ as expected from a fruitbody of a mutualistic fungus (Hobbie *et al.* 2001; 2002) and closer to the ones obtained from non-mycorrhizal tips (tab. 1). Only a few symbiotic organs of ectomycorrhizal species have been investigated regarding $\delta^{15}N$. Regardless of the season fine roots of beech trees had consistently a negative $\delta^{15}N$ between −3.6 and −5.0 ‰, and the values for EM figured between −3,2 and −5,2‰ (Haberer *et al.*, 2007). Therefore, although positive $\delta^{15}N$ values have been reported for fruitbodies (Hobbie *et al.* 2001; 2002), the values for EM are almost identical to those of fine roots (Haberer *et al.*, 2007). $\delta^{15}N$ values of *H. penarius* EM are also in the range of roots, and their $\delta^{15}N$ of fruitbodies are, although rather low in two samples (Tab. 1), as expected for EM fungal fruitbodies (Hobbie *et al.* 2001; 2000). More interestingly, the EM of *Lactarius acris* from the same site as those of *H. penarius* reveal a rather high positive value of 3.35 ‰. But this value resulted as that of *H. penarius* only form a single soil core, indicating, that comprehensive studies of several fruitbodies and of several EM of many species, are necessary until an isotopic contribution to the question, whether *H. penarius* behaves rather as a parasite than a typical mutualist, is possible.

At present, only interpretative speculations on this partnership can be shaped, since many factors may have contributed to the observed relation, e.g. genotypes, source-link relationships, and exceptional soil and climatic conditions.

As previously suggested in other ectomycorrhizal symbioses (Schwacke & Hager, 1992; Salzer *et al.*, 1996), the wide and unspecific host reactions induced by *H. penarius*, could be effective against other microorganisms, in accordance with well known "induced resistance" strategies (Sticher *et al.*, 1997; van Loon *et al.*, 1998).

Further anatomo-physiological investigations on *H. penarius* behaviour in different tip age and seasons, and on its potential ability to induced a non-specific plant resistance to rootlet's parasites are therefore of main importance.

Acknowledgments: Thank to Dr. E. Blaschke, Dr. S. Raidl, Dr. L. Scattolin, Dr. L. Beenken, Miss C. Bubenzer, Mr. E. Marksteiner, Dr. Philomena Bodensteiner, R.Verma and the "Fondazione A. Gini" of the University of Padova for the partial financial support. The research and the soil analyses were mainly supported by the Centro di Ecologia Alpina (TN, Italy), through the "Fondo per i progetti di ricerca della Provincia Autonoma di Trento", "InHumusNat2000" (1587/2004).

References

AGERER R (ED) (1987-1992).Colour atlas of ectomycorrhizae, 1st-6th delivery. Einhorn, Schwäbish Gmünd).

AGERER R (1991). Characterization of ectomycorrhizae. In: Norris JR, Read DJ, Varma AK (eds) Techniques for the study of mycorrhiza. (Methods in microbiology, vol 23) Academic Press, London, pp.25- 73.

AGERER R (1995). Anatomical characteristics of identified ectomycorrhizas: an attempt towards a natural classification. In Varma K, Hock B (eds) mycorrhiza: structure, function, molecular biology and biotechnology. Springer, Berlin, Heidelberg, New York, pp 685-734.

AGERER R (2001). Exploration types of ectomycorrhizal mycelial systems. A proposal to classify mycorrhizal mycelial systems with respect to their ecologically important contact area with the substrate. *Mycorrhiza* 11: 107-114.

AGERER R (2006). Fungal relationship and structural patterns of their ectomycorrhizae. *Mycological Progress* 5 (2): 67-107.

AGERER R, BEENKEN L (1998). *Geastrum fimbriatum* Fr. + *Fagus sylvatica* L. Descr Ectomyc 3: 13-18.

AGERER R, RAMBOLD G (2004-2007). [first posted on 2004-06-01; most recent update: 2007-

05-02]. DEEMY – An Information System for Characterization and Determination of Ectomycorrhizae. www.deemy.de – München, Germany.

AGERER R, WALLER K (1993). Mycorrhizae of *Entoloma saepium*: parasitism or symbiosis? *Mykorrhiza* 3: 145-154.

AHMADJIAN V, PARACER S (1986). Symbiosis – an introduction to biological associations. University Press of New England, Hanover, N.H.

AZUL MA, AGERER R, FREITAS H (2001). "*Quercirhiza ectendotrophica*" + *Quercus suber* L. Descr Ectomyc 5: 67-72.

BAILEY JA (1985). Biology and Molecular Biology of Plant-Pathogen Interactions. Springer-Verlag, Berlin,Germany.

BERG B, GRONBACH E (1989). Piceirhiza gelatinosa (*Hygrophorus olivaceolabus* (Fr.:Fr.)Fr. + *Picea abies*) in Agerer R (ed) Colour Altas of Ectomycorrhiza, plate 30, Einhorn-Verlag, Schwäbisch Gmünd, Germany.

BERG B (1989). Charakterisierung und Vergleich von Ektomykorrhizen gekalkter Fichtenbestände. Diss. Univ. München.

BRAND F (1991). Ektomykorrhizen an Fagus sylvatica. Charakterisierung und Identifizierung, ökologische Kennzeichnung und unsterile Kultivierung. Libri Botanici 2: 1-229.

BRUNDRETT MC (2004). Diversity and classification of mycorrhizal associations. *Biology Reviews* 79: 473-495.

DAWSON TE, MAMBELLI S, PLAMBOECK AH, TEMPLER PH, TU KP (2002). Stable Isotopes in Plant Ecology. *Annual Review of Ecology and Systematics* 33: 507-559.

de BARY A (1887). Comparative morphology and biology of the fungi, mycetozoa and bacteria. Clarendon Press, Oxford.

GEBAUER G, TAYLOR AFS (1999). ^{15}N natural abundace in fruit bodies of different functional groups of fungi in relation to substrate utilization. *New Phytologist* 142: 93-101.

GÖTSCHE D (1972). Mitteilungen der Bundesforschungsamt für Forst- und Holzwirtschaft. Verteilung von Feinwurzeln und Mykorrhizen im Bodenprofil eines Buchen- und Fichtenbestandes im Solling. Hamburg (D).

GRONBACH E(1988). Charakterisierung und Identifizierung von Ektomykorrhizen in einem Fichtenbestand mit Untersuchungen zur Merkmalsvariabilität in sauer beregneten Flächen. Bibliotheca Mycologica 125:1-217.

GRONBACH E (1989a). *Hygrophorus pustulatus* (Pers.) Fr..+ *Picea abies* In Agerer R (ed) Colour Atlas of Ectomycorrhizae, plate 26, Einhorn-Verlag, Schwäbisch Gmünd, Germany.

GRONBACH E (1989b). Piceirhiza glutinosa. In Agerer R (ed) Colour Atlas of Ectomycorrhizae, plate 31, Einhorn-Verlag, Schwäbisch Gmünd.

GRONBACH E, AGERER R (1986). Charakterisierung und Inventur der Fichten-Mykorrhizen im Höglwald und deren Reaktion auf saure Beregnung. Forstwiss Cbl 105: 329-335.

HABERER K, GREBENC T, ALEXOU M, GESSLER A, KRAIGHER H, RENNENBERG H (2007).Effects of longterm freeair ozone fumigation on d15N and total N in *Fagus sylvatica* and associated mycorrhizal fungi. *Plant Biology* 9(2): 242-252.

HAUG I, PRITSCH K (1992). Ectomycorrhizal types of spruce (*Picea abies* (L.) Karst.) in the Black Forest. Kernforschungszentrum Karlsruhe.

HERRE EA, KNOWLTON N, MUELLER UG, REHNER SA (1999). The evolution of mutualismus exploring the paths between conflict and cooperation. Tree (14) 2: 49-53.

HOBBIE EA, MACKO SA, SHUGART HH (1999). Insights into nitrogen and carbon dynamics of ectomycorrhizal and saprotrophic fungi from isotopic evidence. *Oecologia* 118: 353-360.

HOBBIE EA, WEBER NS, TRAPPE JM (2001). Mycorrhizal vs saprotrophic status of fungi: the isotopic evidence. *New Phytologist* 150: 601-610.

HOBBIE EA, WEBER NS, TRAPPE JM, VAN KLINKEN GJ (2002). Using radiocarbon to determine the mycorrhizal status of fungi. *New Phytologist* 156: 129-136.

HOLMGREN, P.K., N.H. HOLMGREN & L.C. BARNETT (1990): Index Herbariorum. Part I. Herbaria of the World. 8th edn. Regnum Vegetabile 120. New York Botanical Garden, New York (http://www.nybg.org/bsci/ih/ih.html).

HORSFALL JG, COWLING EB (ed.) (1980). Plant Disease.- An Advanced Treatise. Vol. 2. How Disease Develops in Populations. Academic Press, New York.

JUMPPONEN A (2001). Dark septate endophytes – are they mycorrhizal? *Mycorrhiza* 11: 207-211.

KOBAYASHI H, HATANO K (2001). A morphological study of the mycorrhiza of *Entoloma clypeatum* f. *hybridum* on *Rosa multiflora*. *Mycoscience* 42: 83-90.

KULDAU GA, YATES IE (2000). Evidence for Fusarium endophytes in cultivated and wild plants. Microbial Endophytes (C.W. Bacon and J.F. White, eds.). Marcel Dekker, New York and Basel.

LU G, CANNON PF, REID A, SIMMONS CM (2004). Diversity and molecular relationships of endophytic *Colletotrichum* isolates from the Iwokrama Forest Reserve. Guyana. *Mycological Research* 108: 53-63.

KOHZU A, YOSHIOLA T, ANDO T, TAKAHSHI M, KOBA K, WADA E (1999). Natural ^{13}C and ^{15}N abundance of field-collected fungi and their ecological implications. *New Phytologist* 144: 323-330.

KÕLJALG U, LARSSON K-H, BARENKOV K, NILSSON RH, ALEXANDER IJ, EBERHARDT U, ERLAND S, HOILAND K, KJOLLER R, LARSSON E, PENNANEN T, SEN R, TAYLOR AFS, TEDERSOO L, VRALSTAD T, URSING BM (2005). UNITE: a database providing web-based methods for the molecular identification of ectomycorrhizal fungi. New Phytologist 166:

1063-1068.

KULDAU GA, YATES IE (2000). Evidence for Fusarium endophytes in cultivated and wild plants. Microbial Endophytes (C.W. Bacon and J.F. White, eds.). Marcel Dekker, New York and Basel.

PETERSON LR, MASSICOTTE HB (2004). Exploring structural definition of mycorrhizas with emphasis on nutrient-exchange interfaces. *Canadian Journal of Botany* 82: 1074-1088.

RODRIGUES KF, SIEBER TN, GRUENIG CR, HOLDENRIEDER O (2004). Characterization of Guignardia mangiferae isolated from tropical plants based on morphology, ISSR-PCR amplifications and ITS1-5.8S-ITS2 sequences. *Mycological Research* 108: 45-52.

SALZER P, HEBE G, REITH A, ZITTERELL-HAID B, STRANSKY H, GASCHLER K, HAGER A (1996). Rapid reactions of spruce cells to elicitors released from the ectomycorrhizal fungus Hebeloma rustiliniforme, and inactivation of these elicitors by extracellular spruce cell enzymes. *Planta* 198: 118-126.

SCHWACKE M, HAGER A (1992). Fungal elicitors induce a transient release of active oxygen species from cultured spruce cells depending on Ca^{2+} and protein-kinase activity. *Planta* 187:136-141.

SIEBER TN (2002). Fungal root endophytes. The Hidden Half (Y. Waisel, A. Eshel, U. Kafkaki, eds.). Marcel Dekker, New York

SMITH SE, READ DJ (1997). Mycorrhizal symbiosis. 2nd ed. Academic Press, Cambridge, UK.

STICHER L, MAUCH-MANI B, METRAUX JP(1997). Systemic acquired resistance. *Annual Review of. Phytopathology* 35: 235-270.

TEDERSOO L,SUVI T, LARSSON E, KÕLJALG U (2006). Diversity and community structure of ectomycorrhizal fungi in a wooded meadow. *Mycological Research* 110:734-748.

TREU R (1990). Charakterisierung und Identifizierung von Ektomykorrhizen aus dem Nationalpark Berchtesgaden. Bibl Mycol 134: 1-196. *Hygrophorus lucorum* Kalchbr. + *Larix.* In: www.deemy.de.

VAN DRIESCHE R. G, BELLOWS T. S. JR., 1996. Biological control. Chapman and Hall, New York. 539 pp.VAN LOON LC, BAKKER PAHM, PIETERSE CM (1998). SYSTEMATIC RESISTANCE INDUCED BY RHIZOSPHERE BACTERIA. *Annual review of Phythopatology* 36: 453-483.

Fig. 1a-c. Habits. - **Fig. 2**. Mantle surface with an extended gelatinous matrix, with soil particles on its surface, clamps and some hyphae with (partially) thickened walls. - **Fig. 3**. Outer mantle layer plectenchymatous with very extended gelatinous matrix, hyphae with thin walls and others with irregularly thickened walls. - **Fig. 4**. Middle mantle layer plectenchymatous with a very extended gelatinous matrix, hyphae with crystals, clamp, and an anastomosis closed by a septum (asterisk); a few hyphae with irregularly thickened walls; part left below shows again a portion of the outer mantle layer. - **Fig. 5**. Middle mantle with very extended gelatinous matrix, rhomboid crystals on hyphae and soil particles incorporated in the matrix. - **Fig. 6a**. Inner mantle layer plectenchymatous with less extended gelatinous matrix, hyphae without clamps, simple septa very infrequent. – **b**. Hyphae with irregularly thickened walls. - **Fig. 7**. Emanating hyphae covered by a gelatinous matrix, with large clamps. *All. figs. from EDM60a*. - **Figs. 8**. Older ectomycorrhizae with emanating hyphae soil particles and sand. - **Fig. 9**. Tip with matrix, and soil particles: the root below the mantle is dark. - **Fig. 10**. Tip with very thick matrix, sand particles on the surface: root is light-brown. - **Fig. 11**. Ectomycorrhizae of *Hygrophorus penarius* together with other types. – **Fig. 12**. Outer mantle, gelatinous matrix (Interference contrast). – **Fig. 13**. Outer mantle: matrix and hyphae with clamps. (Interference contrast). – **Fig. 14**. Emanating hyphae in the matrix: hyphae with thicker walls (Interface contrast). – **Fig. 15**. Emanating hyphae with thicker walls and rest of the matrix (Interface contrast). - **Fig. 16**. Inner mantle with clamp and gelationous matrix (Interface contrast). – **Fig. 17**. Inner mantle (Interface contrast).- **Fig. 18**. Median longitudinal section (phase contrast): intracellular colonization of plant cells (arrow) and plant reaction (m: mantle). - **Fig. 19**. Median longitudinal section (phase contrast): reaction of the plant (m : mantle; v: vessel; w: thick walls; ta: condensed tannins). - **Fig. 20**. Median longitudinal section (phase contrast): intracellular hyphae (arrow) and thicker walls. - **Fig. 21**. Median longitudinal section (phase contrast): intracellular hyphae (arrow). - **Fig. 22**. Medial longitudinal section (phase contrast): partial digestion of cortical cells and prevention of the meristem tissue (older ontogenetical stage). - **Fig. 23**. Median longitudinal section (phase contrast): meristem in the younger ontogenetical stage and intracellular hyphae (me: reduced meristem).

Tab. 1 Stable isotope analysis of beech roots (length = 2 mm) and of different ectomycorrhizae.

Sample (roots/EM)	$\delta^{15}N$	$\delta^{13}C$	$^{14}N:^{15}N$	$^{12}C:^{13}C$
Root (a)	-3.31	-28.09	0.75	49.92
H. penarius (a)	-3.91	-26.84	0.90	42.44
Lactarius acris (a)	**3.35**	-26.04	**2.12**	43.59
Root (b)	-4.00	-28.36	0.73	50.83
H. penarius (b)	-4.00	-26.84	0.86	41.81
Root (c)	-3.56	-28.17	0.75	49.38
H. penarius (c)	-5.38	-26.71	0.86	39.65
Sample (fruitbodies)	$\delta^{15}N$	$\delta^{13}C$	$^{14}N:^{15}N$	$^{12}C:^{13}C$
L. acris	3.54	-23.02	4.09	46.98
H. russula	5.38	-25.71	4.27	37.47
H. penarius 1	1.17	-24.72	4.65	43.40
H. penarius 2	4.77	-22.40	5.56	40.47
H. penarius 3	1.9	-23.29	3.11	38.00

(a), (b), and (c) are different subsamples of a single soil core.

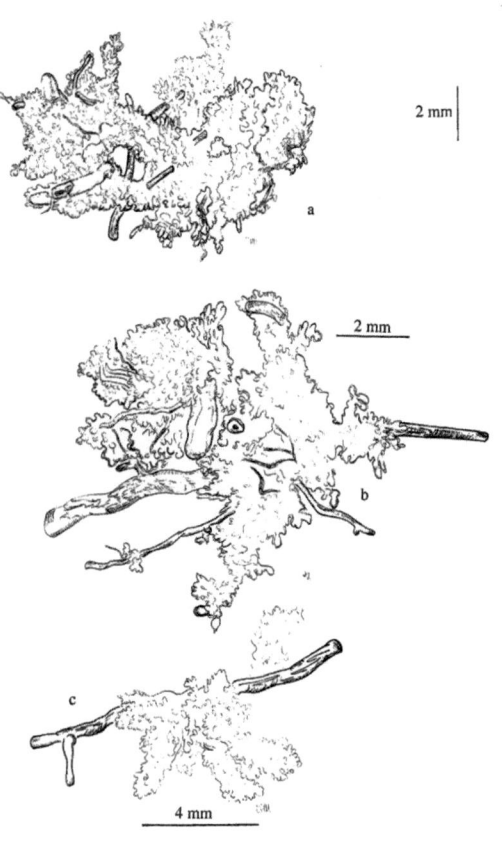

Figs. 1a, 1b, 1c.

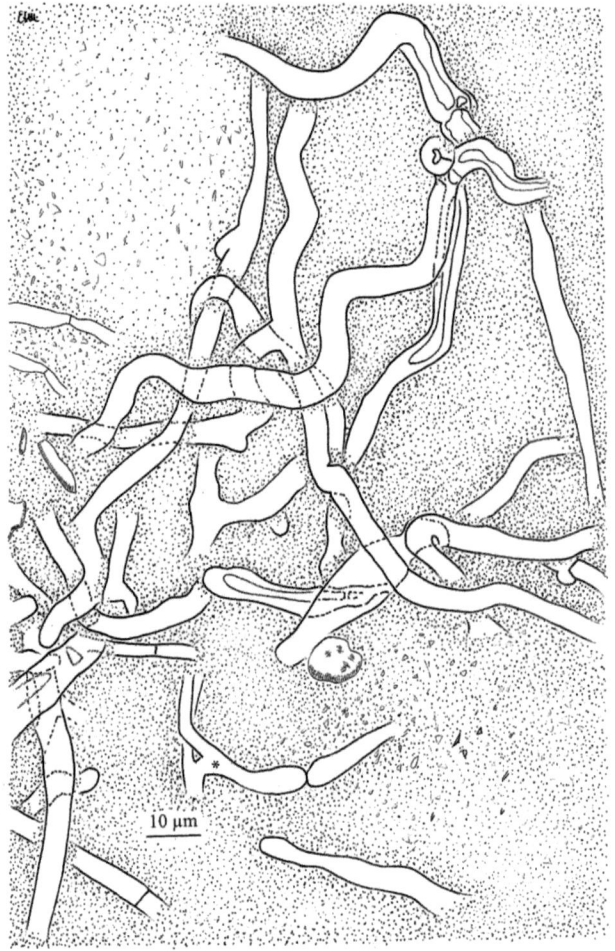

Fig. 2.

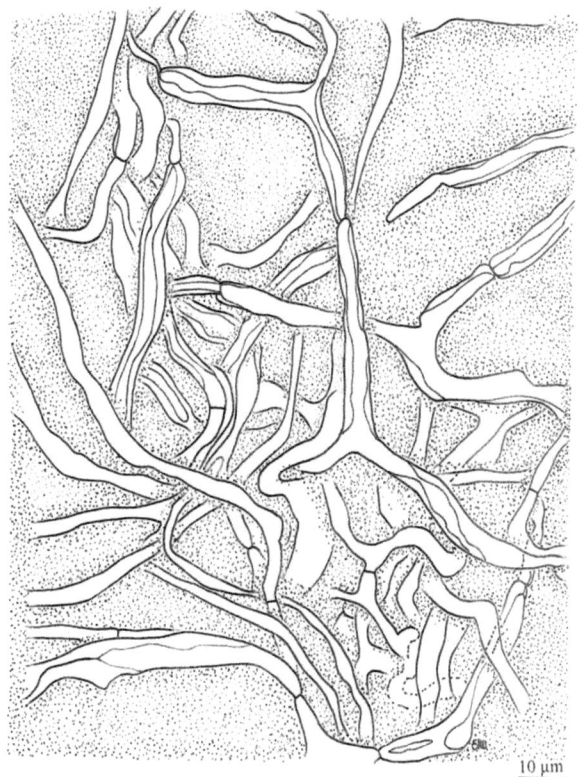

Fig. 3.

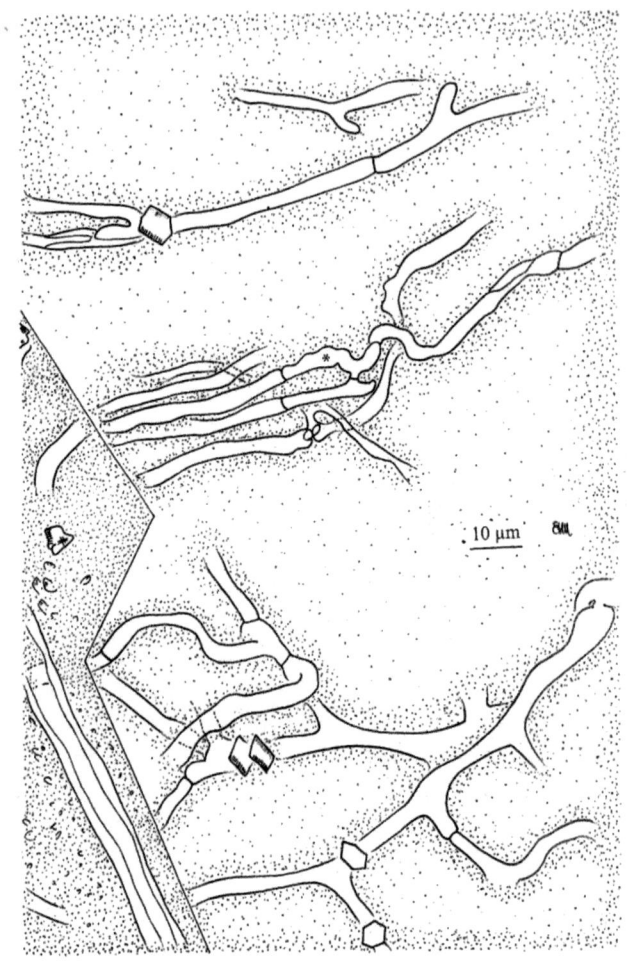

Fig. 4.

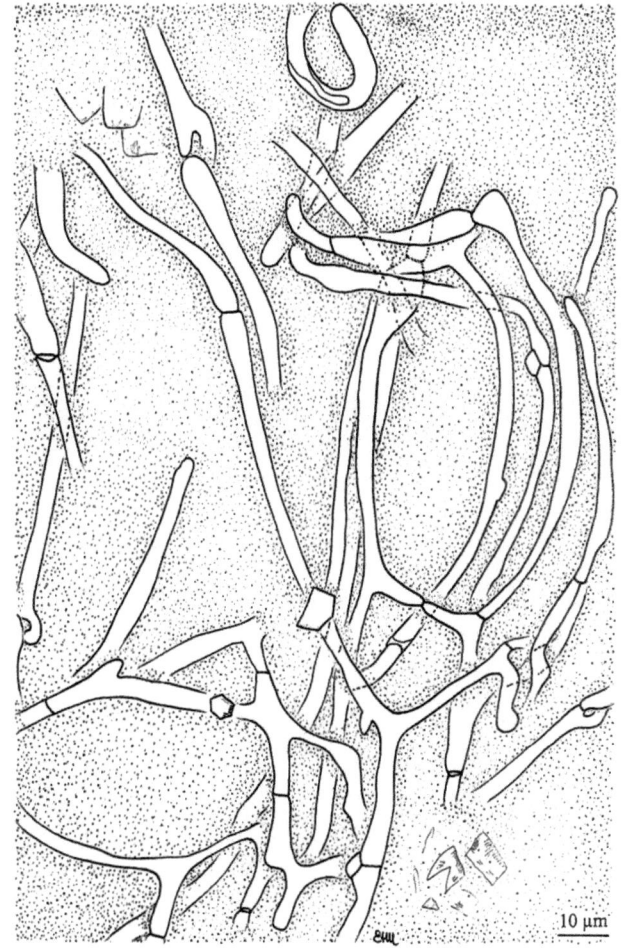

Fig. 5.

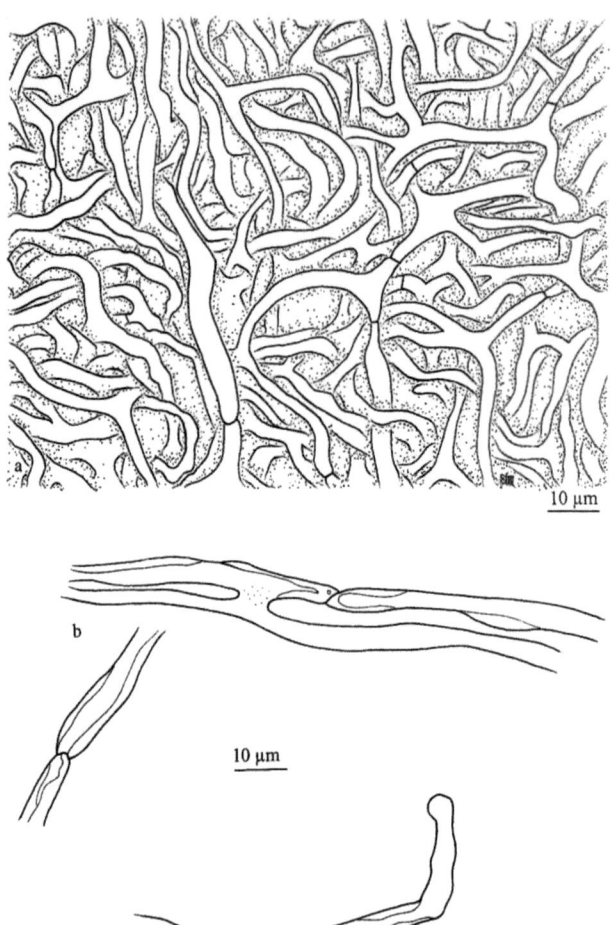

Figs.: 6a, 6b.

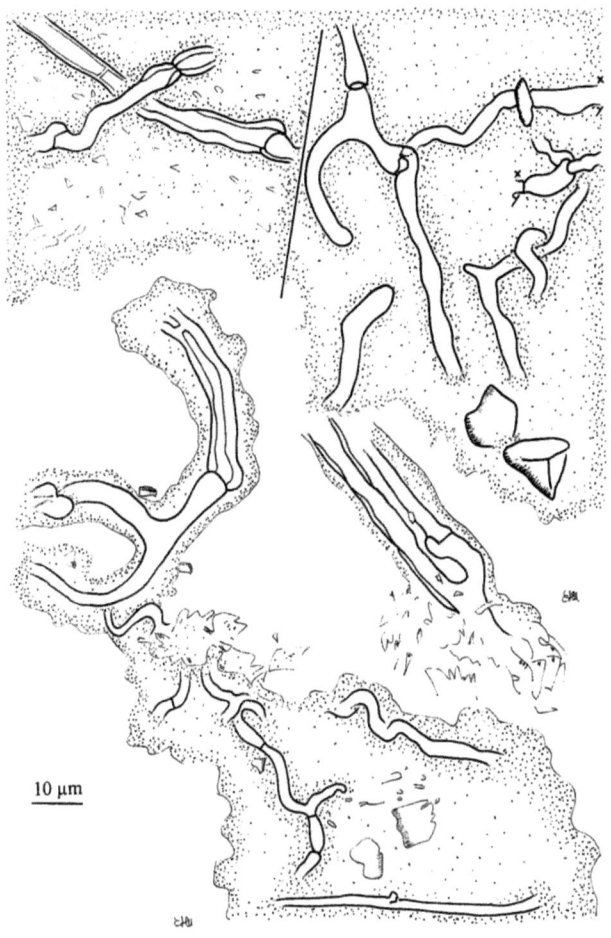

Fig. 7.

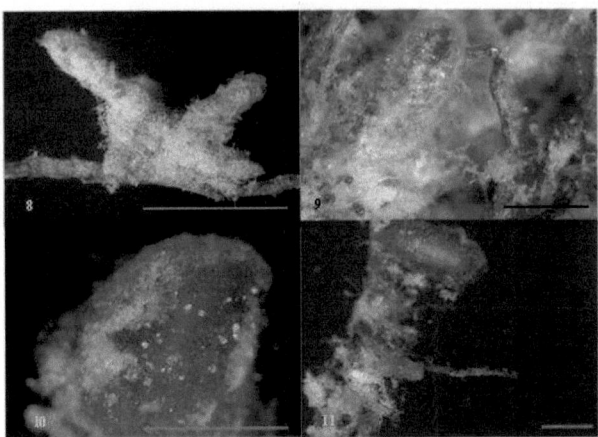

Fig 8: 1 bar = 5 mm. Fig.9: 1 bar = 2 mm. Fig 10: 1 bar = 1 mm. Fig. 5 : 1 bar = 1 mm. Fig. 11: 1 bar = 1 mm.

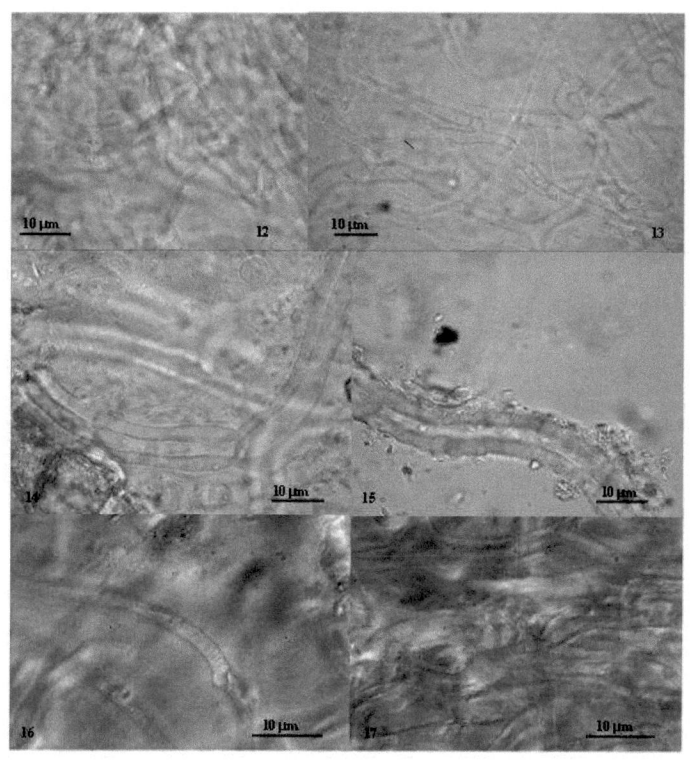

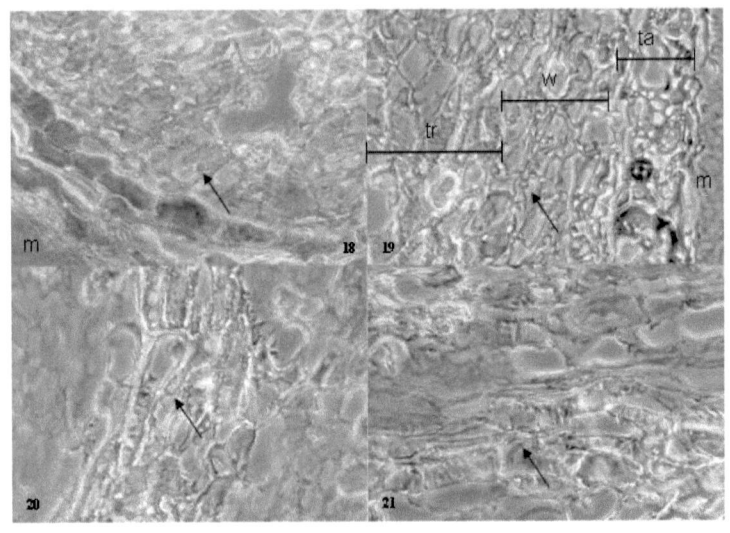

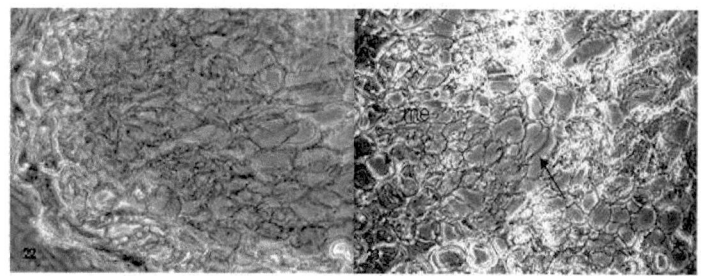

CHAPTER 4

The ectomycorrhizal community structure in beech coppices differing in age and stand features

Erika Di Marino[1,2,3], *Lucio Montecchio*[1], *Linda Scattolin*[2], *Clemens Abs*[4], *Reinhard Agerer*[3]

[1] Università degli Studi di Padova, Dipartimento Territorio e Sistemi Agroforestali, V.le dell'Università, 16, I-35020 Legnaro (PD), Italy.

[2] Centro di Ecologia Alpina, Viote del Monte Bondone, I- 38070 Sopramonte (TN), Italy.

[3] Organismic Biology: Mycology, Department Biology and GeoBio-Center^{LMU}, University of München, Menzinger Str., 67, D-80638, Germany

[4] Fachgebiet Geobotanik, Wissenschaftszentrum Weihenstephan, Technische Universität München,, Am Hochanger 13, D-85354 Freising, Germany

Corresponding author: E. Di Marino; erika.dimarino@unipd.it;

Abstract

The composition and the structure of the ectomycorrhizal (EM) community were investigated in 7 European beech coppices with different rotation (2÷48 years) in northern Italy (Trentino-Südtirol region). To verify whether ectomycorrhization of root tips and the species vary according a reduction of the normal turn applied in these sites, the researches were carried on the spatial and vertical distribution of the EM species. Further ecological factors were considered to study variations according the site features like: slope, pH, C/N, C_{org}, N_{org}, soil moisture, exposure, altitude and bedrock type. The results demonstrated that tips vitality and ectomycorrhization degree did not change significantly either on the same tree, or among trees growing in the same stand, but only between the organic and the mineral soil. The ecological index of richness and evenness attested also only temporal variations, but they were not correlated with the coppice frequency or the slope.

EM species composition didn't reveal a significant correlation with the shoot age but with other site features as the slope and the soil moisture. A particular kind of "resilience" condition was supposed, but further studies are necessary to understand the possible application of the "Short rotation" practices in Beech stands as a sustainable activity, according to the new trends in EU energetic policies.

Key words: EM community, coppice, short rotation, biomass, wood, energy, *Fagus sylvatica*.

1 Introduction

The composition and richness of the ectomycorrhizal (EM) community has high functional significance in forest ecosystems (Peter *et al.*, 2001). In fact, EM structures, morphologically and functionally changing with the fungal species, can acquire and transport water and nutrients from the soil to the plant, buffering short-time water stresses which often predispose the plant to decline or to a early senescence (Skinner & Bowen, 1974; Duddridge *et al.*, 1980; Kammerbauer *et al.*, 1989; Unestam 1991, Smith & Read, 1997; Montecchio *et al.*, 2000; Agerer, 2001, 2006; Werner *et al.*, 2002).

Diversities in EM community structure are a result of site feature, and different sets of species with their site-dependent abundances are possibly able to provide comparable benefits to EM plant communities growing in different sites (Erland & Taylor, 2002; Agerer & Göttlein, 2003; Baier *et al.*, 2006). Tip vitality, degree of mycorrhization, EM richness, species composition and evenness, therefore, might be associated with different environmental variables (Koide *et al.*, 1998; van der Heijden *et al.*, 1999; Scattolin 2007; Scattolin *et al.*, 2008b).

Abiotic (i.e. soil, site and climatic features) and biotic (i.e. parasitic infections) factors can drive to direct or indirect effects on EM species diversity at community level (Bakker *et al.*, 2000; Byrd *et al.*, 2000; Peter *et al.*, 2001; Erland & Taylor 2002; Lilleskov *et al.*, 2002; Shi *et al.*, 2002; Jany *et al.*, 2003; Montecchio *et al.*, 2004; Montecchio 2005), but little is know about the possible role of sylvicultural treatments at forest scale (Buèe *et al.*, 2005; Mosca *et al.*, 2007a, b).

According to both European Union and Italian rules promoting the increase of renewable energetic resources availability (Bernetti *et al.*, 2004) also through "short rotation coppices" policies, the main goal of the research was to verify the effects of both coppice frequency and site features in EM community structure and biodiversity, of healthy Beech forests.

2 Methods

Stand characteristics and sample collection

The investigations were performed in 2005 and 2006 in 7 coeval beech (*Fagus sylvatica* L.) coppice stands growing in the Natural Park Adamello-Brenta (Northern Italy; 5.125.228 ÷ 5.125.666 N, 1.654.361 ÷ 1.654.565 E) selected among the most productive and exploited in the Trentino-Südtirol Region, where the officially fixed coppice frequency is 25 years (Provincia Autonoma di Trento, 2001; 1923; Sboarina & Cescatti, 2004).

To assess the EM spatial distribution, in 2005 four sites differing in coppicing age (coppiced 4 to 47 years before) were selected and coded 1 to 4, respectively, while in 2006 three additional stands (coppiced 2 to 48 years before) were selected and coded 5 to 7. Stands older than 25-years-old were devoted to change to a high forest stand. The dominant humus form was classified according to Jabiol *et al.* (2005), for each stand. It is reported together with other investigated environmental variables in Tab.1.

In each plot, after a phytosanitary survey, 4 healthy stumps, undamaged by climatic events, with fully-developed crown, growing at least 15 m from the nearest EM plant, were randomly selected and coded. According to Scattolin *et al.* (2008a), in June and October 2005 (sites 1÷4) and in June 2006 (sites 5÷7), 12 cylindrical soil cores (Ø 18 mm; 15 cm deep) were collected (100, 150 and 200 cm from the collar, along N, E, S and W directions), stored in plastic pipes at 4 ±1 °C in the dark and used for EM classifications and statistical analyses.

To investigate the EM vertical distribution, in October 2006 two more beech stumps were randomly selected in the sites 1÷5 according to the above reported methods. Four 2.5 x 2.5 cm soil samples including the mineral layer A, were collected along the four cardinal directions 150 cm from the base. Every sample was then vertically divided into 2 equal subsamples and separately preserved (subsamples *a, b*) as above reported. Subsamples *a* were then used for EM classifications and statistical analyses, while from the subsamples *b* the organic horizon O and the mineral horizon A were classified and chemical analyses on N tot, C/N, soil moisture (RH), pH were performed according to official methods (Repubblica Italiana, 1999).

EM classification

Within 12 days from sampling, from each core 10 fully developed rootlets with undamaged apical tips were randomly chosen and carefully cleaned. For the following spatial distribution analyses, the last tip was classified as *not vital* (NV), *vital not-mycorrhizal* (NM) and *vital ectomycorrhizal* (EM). For the vertical distribution analyses, only the mycorrhizal tips were considered, as *not-vital mycorrhizal* (NVM) and *vital ectomycorrhizal* (EM; Montecchio *et al.*, 2004).

Every EM tip was then classified anatomically and morphologically (Agerer 1991; Goodman *et al.*,

1996; Agerer 1987-2006; Brand, 1991a, b; Agerer & Rambold, 2004-2007), and the ones with uncertain classification were submitted to molecular analyses (Tedersoo et al., 2006). DNA extraction, amplification, sequencing were performed according to Benkeen (2004), while the sequence identification level was assigned to species level for similarity of 100%, to genus level 95-99%, to family or ordinal level for < 95%, by means of both Genbank and Unite databases (Kõljalg et al., 2005). Anatomotypes detected with not enough tips to allow the molecular procedure, and the ones with uncertain results were classified by an alphanumerical code (EDMxx).

All specimens were preserved in FEA (formaldehyde 40% : ethyl alcohol 50% : acetic acid 100% : = 5 : 90 : 5, v/v/v) solution and stored in the TeSAF Departmental herbarium, University of Padova.

EM spatial distribution

The studies on spatial distribution were carried out by sampling collections of the first year (June and October 2007, sites 1÷4) and of the second year (June 2006, sites 5÷7). The NV, NM, EM *absolute frequencies* among samples from the same tree, among trees from the same site, and among sites were calculated (Scattolin et al., 2008b) and compared through the Kruskal-Wallis non-parametric test and the χ^2 test (P<0.05, SAS System, SAS Institute, Cary, NC, USA).

As the autocorrelation among sampling points could influence the community structure, the Mantel Test was performed to test the null hypothesis of no relationships among samples from the same tree (McCune & Grace, 2002). The Sørensen similarity index was used to create the similarity matrix: $2a/(2a+b+c)$, where a= number of shared species, b= number of species unique to plot 1 and c= number of species unique to plot 2 (Izzo et al., 2005). The Mantel Test ($P<0.01$; permutations=10000; R program, www.r-project.org vers. 2.5.1.) was used to compare EM species dissimilarity and linear distance matrixes between sampling points belonging to the same plant.

To avoid seasonal effects in EM community structure, the EM absolute frequencies recorded in sites 1÷4 in the two sampling periods were assembled (Baier et al., 2006).

Relations among EM absolute *frequency* in each sample and coppice frequency (age), slope gradient, altitude, exposure, humus and bedrock type, were analysed by means of multivariate techniques : Detrended Correspondence (DCA; Hill & Gauch, 1980) and Canonical Correspondence Analyses (CCA; Hill 1979) were carried out (McCune & Mefford, 1999; PC-ORD^{TM} vers. 5, MjM, Oregon) applying a power transformation (power = 0.50, square root) to reduce the number of interactions among the involved variables. During the preliminary analyses, the ecological factors investigated that were correlated with others, were excluded progressively, to prevent interpretation based on autocorrelation.

To assess the type of distribution of the EM communities, the data belonging to the EM relative abundances in all plots were submitted to the Morisita's Index of Dispersion (MID; Morisita 1959; Sokal & Rohlf, 1981), due to its independency by the number of samples, density of the population studied, and sampling size (Krebs 1989).

Biodiversity parameters for each EM community in all sites and samples were calculated by means of the Evenness and the Shannon-Weaver (1949) Index of Diversity using the *absolute frequency* of the species.

EM vertical distribution

EM and NVM *absolute frequencies* among samples from the same tree, among trees from the same site, among sites and between O and A soil horizons, were calculated and compared through the Kruskal-Wallis non-parametric test and the χ^2 test as above reported.

Lacking significant differences among sampling directions (Mantel Test; Mc-Cune and Grace 2002), EM and NVM data from each tree were gathered and the *relative abundance* (Σ EM/cm^3 soil volume and Σ NVM/cm^3 soil volume) were calculated and related to coppice age, slope gradient, altitude, humus, bedrock type, soil horizons (O, A), pH, moisture, C_{org}, N_{tot}, C/N by means of DCA and CCA, as above reported.

Morisita's Index of Dispersion, Evenness and the Shannon-Weaver (1949) Index of Diversity using the *absolute frequency* of the species were calculated, too.

3 Results

EM community structure: spatial distribution

Anatomical and molecular investigations revealed a total of 46 anatomotypes (Tab. 6). Among them, 3 were assigned to family or ordinal level (*Boletales, Sebacinaceae, Thelephorales*), 15 to a genus (*Amphinema* sp., *Boletus* sp., *Cortinarius* sp., *Craterellus* sp., *Hydnum* sp., *Hygrophorus* sp., *Inocybe* sp., *Laccaria* sp., *Lactarius* sp., *Ramaria* sp., *Sebacina* sp., *Tomentella* sp.), 17 to a species [*Byssocorticium atrovirens* (Fr.) Bondartsev and Singer ex Singer, *Cenococcum geophilum* Fr., *Cortinarius bolaris* (Pers.) Fr., *Cortinarius cinnabarinus* Fr., *Cortinarius ionochlorus* Maire, *Genea hispidula* Berk. ex Tul. and C. Tul., *Hygrophorus penarius* Fr., *Lactarius acris* (Bolton) Gray, *L. pallidus* W. Sounders and W. G. Sm., *L. rubrocinctus* Fr., *L. subdulcis* (Bull.) Gray, *L. vellereus* (Fr.) Fr., *Piloderma croceum* J. Erikss. and Hjortstam, *Ramaria aurea* (Schaeff) Quél, *Russula illota* Romagn., *R. mairei* Singer, *Tricholoma acerbum* (Bull.) Vent., *T. sciodes* (Pers.) C. Martin], 8 were previously described in detail on *Fagus sylvatica* [*Fagirhiza cystidiophora* (Brand 1991a), *F. fusca* (Brand 1991a), *F. lanata* (Brand 1991a), *F. oleifera* (Brand 1991a), *F. pallida* (Brand 1991a), *F. setifera* (Brand 1991a), *F. spinulosa* (Brand 1991a), *F. vermiculiformis* (Jakucs 1998)], three were studied in all their features and their description by the first author is in progress for following

(*Fagirhiza byssoporioides*, *F. entolomoides*, *F. stellata*), while one remained unknown (Tab. 7). Both in June and October, *C. geophilum* was the dominant species (19.2 %). *C. geophilum*, *L. pallidus*, *L. vellereus* and *Hydnum* sp. altogether represented 60.8% of the whole EM species, while 8 *Thelephoraceae* represented 17.4%. The absolute abundance of the anatomotypes is reported in the table 2.

In sites 5÷7 the investigations demonstrated the presence of 36 anatomotypes, the most of them previously detected in sites 1÷4. Additional anatomotypes were *H. penarius*, *Fagirhiza arachnoidea* (Brand 1991a), a *Ramaria* sp. and a species belonging to *Thelephorales* (Tab. 2).

No EM spatial autocorrelation was found among samples from the same tree (Mantel test; P<0.05). The performed analyses demonstrated that NV, NM and EM do not differ significantly among samples belonging to the same tree (different directions and distances from the collar), among trees belonging to the same site, and among sites (P<0.05). Furthermore, the same parameters did not differ significantly with both coppice age and bedrock type in all the 7 investigated sites (P<0.05).

The DCA performed on the data-set collected in the 2005 (not showed here) demonstrated that the length of the main gradient was less than 2, and the total "inertia" (variance) in the species was 0.7840. This analyses performed to evaluate a gradient in the EM spatial distribution with the bedrock type, shoot age, humus type, site slope gradient, altitude and exposure gave insignificant results.

CCA (Tab. 8, Fig.1) revealed that the species were significantly correlated to the slope gradient and the humus form, while the bedrock type and the coppice age gave statistically insignificant effects (total inertia = 0.8079; eigenvalue of the 1st and 2nd axis 0.089 and 0.064, respectively). The correlation measured on the first axis ["intraset correlations" (ter Braak 1986)] showed that the species distribution was highly correlated with the slope (0.650), and that the second most important factor was the shoot age (0.312). A negative correlation revealed with the humus form (-0.706), while a opposite relation was found between slope gradient and humus form. The analyses on EM community in the site 1÷4 demonstrated also, that the species distribution was significantly related to the slope (Fig.1). *Tomentella2* sp., e. g., demonstrated to prefer the plots with steeper slope, while *Amphinema* sp., *Tomentella1* sp. and *Entoloma2* sp. were primarily associated to plots with Amphimull/Dysmull humus type and flatter slope.

Very similar results were obtained with the data set of the sampling in the 5÷7 sites (June 2006). No gradient was found using the same ecological factors (without power transformation, not shown here; total inertia 1.2647, gradient > 1). The results of the CCA performed on the data-set of the calcareous plots studied in 2006 (Tab. 9, Fig.2) confirmed also in this case, that the slope resulted to be the first important ecological factor involving the EM composition. It was revealed an opposite relation between slope gradient and shoot age (total inertia 1.0933; eigenvalue of the first and second axes 0.131 and 0.092, respectively), too. The intraset correlations (ter Braak 1986) showed that the species distribution was more related to the slope (0.503), with a negative correlation with the shoot age (-0. 890). One group of species, including *Fagirhiza arachnoidea, Tomentella*2 sp., *Laccaria* sp., *Tricholoma acerbum*) resulted to be associated to plots with high slope values and younger coppice. Vice versa, *Ramaria*2 sp., *Ramaria*1 sp., and *Thelephorales*1 sp. were mainly present in plots with flatter slope and older coppices.

The MID index showed that the community structure was formed by species always aggregated in the three different collections (MID>1), when the test F was significant (Fo> 1.45): in the first collection (June 2005) this index was always significant (Fo> 1.45), in the second collection (October 2005) this index was significant for only 23 species compared to the total of 46 species, while for the last collection (June 2006) was significant for 19 species compared to the total 29 species. The EM richness and evenness were different among sites regarding spatial distribution but it was never correlated with coppice age or slope (Tab. 3).

EM community structure: vertical distribution

Forty-three anatomotypes were distinguished from 45.946 root tips (Tab. 6) within the plots 1÷5. Among them, 7 were assigned to family or order level (*Boletales, Sebacinaceae, Thelephoraceae*), 10 to genus (*Boletus* sp., *Cortinarius* sp., *Hygrophorus* sp., *Laccaria* sp., *Ramaria* sp., *Sebacina* sp., *Tomentella* sp.), 14 to species [*Byssocorticium atrovirens* (Fr.) Bondartev and Singer ex Singer, *Cenococcum geophilum* Fr., *Cortinarius bolaris* (Pers.) Fr., *Cortinarius cinnabarinus* Fr., *Cortinarius infractus* Berk., *Cortinarius inochlorus* Maire, *Genea hispidula* Berk. ex Tul. and C. Tul., *Lactarius acris* (Bolton) Gray, *Lactarius pallidus* W. Sounders and W. G. Sm., *Lactarius subdulcis* (Bull.) Gray, *Ramaria aurea* (Schaeff) Quél, *Russula mairei* Singer, *Tricholoma acerbum* (Bull.) Vent., *Tricholoma sciodes* (Pers.) C. Martin] and 10 [*Fagirhiza arachnoidea* (Brand 1991a), *F. byssoporioides, F. entolomoides, F. fusca* (Brand 1991a), *F. lanata* (Brand 1991a), *F. oleifera* (Brand 1991a), *F. pallida* (Brand 1991a), *F. setifera* (Brand 1991a), *F. spinulosa* (Brand 1991a), *F. stellata*] to anatomotypes not identified at species level, while 2 remained unknown (Tab. 6).

With a proportion of 29.2%, *Cenococcum geophilum* was the dominant species regarding the total amount of ECM. *C. geophilum, Lactarius pallidus, Cortinarius cinnabarinus* and *Hygrophorus*1 sp. represented 42.6% of all the EM species within the plots, while the genus *Lactarius* alone stand for 12.2% of the EM population, and the genus *Cortinarius* 11.4%. Furthermore the ten Thelephoraceae (*F. fusca, F. lanata, F. spinulosa; F. stellata, Tomentella* spp.) represented 11.5% of the total number of EM tips. In total, 46 different anatomotypes were found in the 7 stands (Tab. 6). The performed analyses demonstrated that among samples collected from the same tree (different directions), NVM and EM never differed significantly within the sites (Kruskal-Wallis Test,

P<0.05). Also no significant differences for the vitality of the EM were observed between the two different soil horizons (Kruskal-Wallis Test, P<0.05), in contrast to the NVM amount, changing significantly with the soil layer because of the higher abundance of EM in the upper one (Kruskal-Wallis Test , P< 0.05; χ^2 test, P<0.05; Tab. 5).

The species distribution mainly depends on soil moisture (DCA diagram, total inertia 1.2647 and length of gradient > 1; Fig. 3). On the first axis we found a positive trend with the sample moisture and a negative trend with the humus form. The species with more moisture in the samples are concentrated on the right of the diagram, and on the left the species with lower moisture in the sample and prefered plots with a humus form between Amphimull and Dysmull. The second less important trend (on the second axis) is related to the age of the coppice. From up to down the diagram shows a trend with shoot age: above the species of older sites, and below the species of younger sites. In the CCA on the data set with assembled directions (Tab. 9, Fig.4), the species were segregated into groups depending mainly on soil moisture. Total inertia in the species data was 1.0933 and the eigenvalue of the first and second axes were 0.131 and 0.092, respectively.

The correlation measured on the first axis ["intraset correlations" (ter Braak 1986)] showed that the species distribution was more correlated to soil moisture with a value of 0.447. The second most important factor was the slope with a value of 0.276, and the third factor was pH with a coefficient of 0.166. Negative correlations mainly with humus form with a coefficient of -0.684 were found. The C/N ratio showed also a negative correlation (-0.458), united with altitude (-0.104) and age (shoot age, -0.083). An opposite relation was found between soil moisture and the humus form.

One species group, counting *Tomentella*2 sp., *Cortinarius*1 sp. and *Fagirhiza arachnoidea*, showed a clear preference for plots with higher soil moisture.

A second group, including *Ramaria*2 sp. and *Thelphoraceae*2 were primarily found with Amphimull/Dysmull humus form and with lower soil moisture.

The Shannon-Weaver index showed that the EM richness and the evenness were different in the sites, but not correlated with the age (shoot age) of the coppices (Tab. 3). Only when the last cut was applied 5 years ago richness and evenness are higher in the mineral horizon. The MID index was always > 1, showing a community structure with 39 species aggregated in the sites when the test F was significat (Fo >1.32; Tab. 5).

4 Discussions

The research was performed in Beech coppices of different age to verify the role of both coppice age and site features in EM community structure and biodiversity.

The achieved results demonstrated that along a wide coppice age gradient (2 to 48 years, with 25 years being the rule), tips' vitality and mycorrhization change only in the vertical distribution with a major abundance of EM not vital in the organic soil layers (Tab. 5), as reported also in previously investigations (Baier et al., 2006). The ecological indexes attested that the richness and evenness varied only on the temporal scale (related to the different collections), but they were not correlated with the coppice age or the slope (Tab. 3), partly confirming available information from clear-cutting and thinning experiments (Buée et al., 2005; Cline et al., 2005; Mosca et al., 2007a), and explainable with an hypothetical resilience, as an "adaptive diversity".

The multivariate analyses based on the ordination techniques, revealed that the slope was the mort important factor explaining the EM community in the spatial distribution in all the investigated coppices (Tabs.7, 8; Figs. 2, 3). The vertical distribution significantly was correlated only with the moisture (Tab. 9, Fig.4), probably due to the higher organic accumulation and moisture availability in the upper soil of the down slope, as reported by other authors (Binkley & Vitousek, 1989; Tateno et al., 2004; Scattolin et al., 2008a).

The other ecological variables like, altitude, exposure, humus, bedrock type, soil horizons (O, A), pH, C_{org}, and N_{tot} never acted as significant driving factors, due to their high correlations with the main factor showed in the ordination diagram and measured in the preliminary elaboration. In addition, the coppice age didn't explain alone the distribution of the EM species (Tabs. 7, 8, 9).

According to previous results valid for different plant species (Grogan et al., 2000; Horton & Bruns, 2001; Taylor 2002; Montecchio et al., 2004; Mosca et al., 2007a; Scattolin et al., 2008b), the EM community resulted to be characterized by few abundant species and many with a significantly lower abundance. In total 46, anatomotypes were observed, with a high proportion of Thelephoroid and Cortinareaceous fungi. This composition is well-known thanks to recent researches, which showed the evidence of EM frequently formed by the Basidiomycote order Thelephorales (Jakucs et al., 2005; Kõljalg et al., 2000; 2001; 2002). The presence of *Cortinarius* species was also discussed, because these species appeared to be less dominant as mycelia than as root tips (Kjøller 2006), instead. *Cenococcum geophilum* was the most frequently detected species in each site and in each period, both in dolomitic and calcareous sites, probably due to its amplitude and antagonistic behaviour versus other EM fungi (Jany et al., 2003; Koide et al., 2005).

Certainly due to their high abundance, the most frequent EM species revealed to have an aggregated distribution, probably due to micro-scale effects (i.e. interactions among species) able to prevail on macro-scale features (i.e. humus and bedrock type), as reported by other authors (Bruns 1995; Toljander et al., 2006; Gebhardt et al., 2007).

The results of the present study are not exhaustive, but they demonstrated that, in respect to the effect of non modifiable features as slope and soil moisture, the coppice age (2 to 48 years) in healthy Beeches doesn't have a primary, significant effect on the EM richness and the community structure. Unfortunately this is the first research on this topic, and the hypothesis of a long-term resilience acquired by an EM community living in old root system subjected to periodical thinning should be demonstrated. Moreover the application of the EM community like a index to understand the assessment of topsoil properties and the litter dynamic will need further research effort. It will be necessary to understand the possibility of a reduction of the organic layers already reported (Buckley 1992) and the lack of significant differences about the EM distribution, considering the organic and the mineral layers as a probable indirect effect of the coppicing.

For assessing ecosystem resilience within the context of the global change, the identification of the ecological features determining this "adaptive diversity" in EM communities, will have more and more importance (Dahlberg 2001).

Taking into account the stability of the EM community as a possible indicator of plant health status (Wargo 1988; Fellner & Caisovà, 1994; Causin *et al.*, 1996; Montecchio *et al.*, 2004; Mosca *et al.*, 2007a), "Short rotation" practices in Beech forests could be considered a sustainable activity, according to the new trends in EU energetic policies, aimed to promote the increase of renewable energetic resources availability (Cutini 2001). From this point of view, new guidelines could be provided for the sylviculture management. Further investigations to verify if and how a high and repeated coppice frequency can drive to irreversible alterations in EM biodiversity are needed.

Acknowledgements

The research was supported by the Centro di Ecologia Alpina (TN, Italy), through the "Fondo per i progetti di ricerca della Provincia autonoma di Trento", "InHumusNat2000" (1587/2004), and by the Fondazione "Ing. A. Gini" (Università di Padova).

References

AGERER R (1991). Characterization of ectomycorrhizae. In: NORRIS JR, READ DJ, VARMA AK, (eds) Techniques for the Study of Mycorrhiza. UK Academic Press, London, pp 25-73.

AGERER R (2001). Exploration types of ectomycorrhizae: a proposal to classify ectomycorrhizal mycelial systems according to their patterns of differentiation and putative ecological importance. *Mycorrhiza* 11: 107-114.

AGERER R (2006). Fungal relationships and structural identity of their ectomycorrhizae. *Mycological Progress* 5(2): 67-107.

AGERER R (ed) (1987-2006). Colour atlas of ectomycorrhizae. 1-12th delivery, Einhorn, Schwäbisch Gmünd, D

AGERER R, RAMBOLD G (2004–2007) [first posted on 2004-06-01; most recent update: 2007-05-02] DEEMY – An Information System for Characterization and Determination of Ectomycorrhizae. www.deemy.de - München, D

AGERER R, GÖTTLEIN A (2003). Correlation between projecton area of ectomycorrhizae and H_2O extractable nutrients in organic soil layers. *Mycological Progress* 2(1):45-52.

BAIER R, INGENHAAG J, BLASCHKE H, GÖTTLEIN A, AGERER R (2006). Vertical distribution of an ectomycorrhizal community in upper soil horizons of a young Norway spruce (*Picea abies* [L.] Karst.) stand of the Bavarian Limestone Alps. *Mycorrhiza* 16(6): 197-206.

BAKKER MR, GARBAYE J, NYS C (2000). Effects of liming on the ectomycorrhizal status of oak. *Forest Ecology and Management* 126(2):121-131.

BENKEEN L (2004). Die Gattung *Russula*. Untersuchungen zi ihrer Systematik anhand von Ektomykorrizen. Dissertation zur Erlangung des Grades eines Doktors der Naturwissenschaften der Fakultät Biologie der Ludwig-Maximilians-Universität München.

BERNETTI I, FAGARAZZI C, FRATINI R (2004). A methodology to analyse the potential development of biomass-energy sector: an application in Tuscany. *Forest Policy and Economics* 6: 415-432.

BINKLEY, D., VITOUSEK, P. 1989. Soil nutrient availability. In Plant physiological ecology. Pearcy, R. W.; Ehleringer, J.; Moony, H. A.; Rundel, P. W. Eds, Chapman and Hall, London, 75-96.

BRAND F (1991a). Ektomycorrhizen an *Fagus sylvatica*. Charakterisierung und Identifizierung, ökologische Kennzeichnung und unsterile Kultivierung. Libri Botanici, IHW-Verlag : 1-229.

BRAND F (1991b). *Russula mairei*. In: Agerer R (ed) Colour atlas of ectomycorrhizae, plate 65. Einhorn, Schwäbisch Gmünd.

BRUNS TD (1995). Thoughts on the processes that maintain local species diversity of ectomycorrhizal fungi. *Plant Soil* 172: 17-27.

BUCKLEY GP (1992). Ecology and Management of coppice woodland (Hardcover).

BUÉE M, VAIRRELES D, GARBAYE J (2005). Year-round monitoring of diversity and potential metabolic activity of ectomycorrhizal community in a beech (*Fagus sylvatica* L.) forest subjected to two thinning regimes. *Mycorrhiza,* 15: 235-245.

BYRD KB, PARKER VT, VOGLER DR, CULLINGS KW (2000). The influence of clear-cutting on ectomycorrhizal fungus diversity in a lodgepole pine (*Pinus contorta*) stand, Yellowstone National Park, Wyoming, and Gallatin National Forest, Montana. *Canadian Journal of Botany* 78: 149-156.

CAUSIN R, MONTECCHIO L, MUTTO ACCORDI S (1996). Probability of ectomycorrhizal infection in a declining stand of common oak. *Annales des Sciences Forestieres* 53: 743-752.

CLINE ET, AMMIRATI JF, EDMONDS FRL (2005). Does proximity to mature trees influence ectomycorrhizal fungus communities of Douglas-fir seedling? *New Phytologist* 166: 993-1009.

CUTINI A (2001). New management options in chestnut coppices: an evaluation on ecological bases. Forest Ecology and Management 141(3): 165-174.

DAHLBERG A (2001). Community ecology of ectomycorrhizal fungi: an advancing interdisciplinary field. *New Phytologist* 150: 555-562.

DUDDRIDGE JA, MALIBARI A, READ DJ (1980). Structure and function of mycorrhizal rhizomorphs with special reference to their role in water transport. *Nature* 287: 834–846.

ERLAND S, TAYLOR AFS (2002). Diversity of ectomycorrhizal fungal communities in relation to the abiotic environment. In var der Heijden MGA, Sanders I (eds) Mycorrhizal ecology. Ecological Studies. Springer, Berlin Heidelberg New York, 163-200.

FELLNER R, CAISOVÀ V (1994). In: Proceeding of Environmental constraints and Oak: ecological and physiological aspects. 29 August-1 September 1994, Nancy France p.142.

GEBHARDT S, NEUBERT J, WÖLLECKE B, MÜNZENBERGER B, HÜTTL RF (2007). Ectomycorrhiza communities of red oak (*Quercus rubra* L.) of different age in the Lusatian lignite mining distict, East Germany. *Mycorrhiza* 17: 279-290.

GOODMAN DM, DURALL DM, TROFYMOW JA, BERCH SM (eds) (1996). A manual of concise descriptions of north american ectomycorrhizae: including microscopic and molecular characterization. *Mycologue Publications and the Canada-BC Forest Resource Development*

Agreement, *Pacific Forestry Centre*, Victoria, B.C.

GROGAN P, BAAR J, BRUNS TD (2000). Below-ground ectomycorrhizal community structure in a recently burned bishop pine forest. *Journal of Ecology* 88: 1051-1062.

HILL MO (1979). DECORAN-A Fortran program for detrended corrispondence analysis and reciprocal averaging. Ecology and Systematics, Cornell University, Ithaca New York.

HILL MO, GAUCH HG 1980.Detrended correspondence analysis: an improved ordination technique. *Vegetatio* 42: 47-58.

HORTON TR, BRUNS TD (2001). The molecular revolution in ectomycorrhizal ecology: pecking into black-box. *Molecular Ecology* 10: 1855-1832.

IZZO A, AGBOWO J, BRUNS TD (2005). Detection of plot-level changes in ectomycorrhizal communities across years in an old-growth mixed-conifer forest. *New Phytolosist 166:* 619–630.

JABIOL B, BRETHES A, PONGE JF, TOUTAIN F, BRUN JJ (eds) (1995) L'humus sous toutes ses formes. ENGREF, Nancy, F

JAKUCS E (1998). "*Fagirhiza vermiculiformis* + *Fagus sylvatica* L." in *Descriptions of Ectomycorrhizae* 3: 7-11.

JAKUCS E, KOVACS GM, AGERER R, ROMSICS C, ERÖS-HONTI Z (2005). Morphological-anatomical characterization and molecular identification of *Tomentella stuposa* ectomycorrhizae and related anatomotypes. *Mycorrhiza* 15: 247-258.

JANY JL, MARTIN F, GARBAYE J (2003). Respiration activity of ectomycorrhizas from *Cenococcum geophilum* and *Lactarius* sp. in relation to soil water potential in five beech forests. *Plant and Soil* 255: 487-494.

KAMMERBAUER H, AGERER R, SANDERMANN H (1989). Studies on ectomycorrhiza XXII. Mycorrhizal rhizomorphs of *Telephora terrestris* and *Pisolithus tinctorius* in association with Norway spruce *(Picea abies):*formation in vitro and translocation of phosphate *Trees*:78-84.

KOIDE RT, XU B, SHARDA J, LEKBERG Y, OSTIGUY N (2005). Evidence of species interactions within an ectomycorrhizal fungal community. *New Phytologist* 165:305-316.

KOIDE RT, SUOMI L, STEVENS CM, McCORMICK L (1998). Interaction between needles of *Pinus resinosa* and ectomycorrhizal fungi. *New Phytologist* 140: 539-547.

KÕLJALG U, DAHLBERG A, TAYLOR AFS, LARSSON E, HALLENBERG N, STENLID J, LARSSON KH, FRANSSON PM, KÅRÉN, JONSSON L (2000). Diversity and abundance of resupinate thelephoroid fungi as ectomycorrhizal symbionts in Swedish boreal forests. *Molecular Ecology* 9: 1985-1996.

KÕLJALG U, JAKUCS E, BOKA K, AGERER R (2001). Three ectomycorrhiza with cystidia formed by different *Tomentella* species as revealed by rDNA ITS sequences and anatomical characteristics. *Folia Cryptog. Estonia Facs.* 38: 27-39.

KÕLJALG U, TAMMI H, TIMONEN S, AGERER R, SEN R (2002). ITA rDNA sequence-based phylogenetic analysis of *Tomentellopsis* species from boreal and temperate forests, and the identification of pink-type ectomycorrhizas. *Mycological Progress* 1: 81-92.

KÕLJALG U, LARSSON K-H, ABARENKOV K, NILSSON RH, ALEXANDER IJ, EBERHARDT U, ERLAND S, HOILAND K, KJOLLER R, LARSSON E, PENNANEN T, SEN R, TAYLOR AFS, TEDERSOO L, VRALSTAD T, URSING BM. (2005) UNITE: a database providing web-based methods for the molecular identification of ectomycorrhizal fungi. *New Phytologist* 166: 1063-1068.

KJØLLER R (2006). Disproortionate abundance between ectomycorrhizal root tips and their associated mycelia. *FEMS Mycrobiology Ecology* (58)2: 214-224.

KREBS CJ (1989). Ecological Methodology. Harper Collins: New York.

LILLESKOV EA, FAHEY TJ, HORTON TR, LOVETT GM (2002). Belowground Ectomycorrhizal Fungal Community Change Over a Nitrogen Deposition Gradient in Alaska. *Ecology* 83 (1): 104-115.

McCUNE B, GRACE JB (2002). Analysis of Ecological Communities. MjM Software, Gleneden Beach, Oregon.

McCUNE B, MEFFORD MJ (1999). PC-ORD^{TM} . Multivariate Analysis of Ecological Data. Version 5 for Windows, MjM Software Gleneden Beach Oregon.

MONTECCHIO L (2005). Damping-Off of Beech Seedlings Caused by *Fusarium avenaceum* in Italy. *Plant Disease* 89:1014.

MONTECCHIO L, CAUSIN R, MUTTO ACCORDI S (2000). Ectomycorrhizae and their involvement in forest decline. In: Ragazzi A., Dellavalle I. (eds.). Decline of oak species in Italy. Problems and perspectives. Accademia di Scienze Forestali, Firenze. pp. 115-128.

MONTECCHIO L, CAUSIN R, ROSSI S, MUTTO ACCORDI S (2004). Changes in ectomycorrhizal diversity in a declining *Quercus ilex* coastal forest. *Phytopathologia Mediterranea* 43: 26-34.

MORISITA M (1959). Measuring of interspecific association and similarity between communities. Mem Fac Sci. Kyoto University, Ser.E (Biol.) 3: pp. 65 80.

MOSCA E, MONTECCHIO L, SELLA L, GARBAYE J (2007a).Short-term effect of removing tree competition on the ectomycorrhizal status of a declining pedunculate oak forest (*Quercus robur* L.). *Forest Ecology and Management* 244: 129-140.

MOSCA E, MONTECCHIO L, SCATTOLIN L, GARBAYE J (2007b). Enzymatic activities of three ectomycorrhizal types of *Quercus robur* L. in relation to tree decline and thinning. *Soil Biology and Biochemistry* 39: 2897-2904.

PETER M, AYER F, EGLI S (2001). Nitrogen addition in a Norway spruce stand altered macromycete sporocarp production and below-ground ectomycorrhizal species composition. *New*

*Phytologist 14*9: 311-325.

PROVINCIA AUTONOMA DI TRENTO. Servizio foreste. Prescrizione di massima e di polizia forestale ai sensi degli artt. 8,9, 10 del RD. 1923/3267 – per i boschi e terreni di montagna sottoposti a vincolo nella provincia di Trento – approvate con decreto del Min. Agricoltura e Foreste del 7.2.1930.

PROVINCIA AUTONOMA DI TRENTO (2001). I dati della pianificazione forestale aggiornati al 31/12/2000. Servizio Foreste-Sistema Informativo Ambiente e Territorio. CD-ROM, Trento, I

R PROGRAM. The R project for statistical Computing. www.r-project.org vers. 2.5.1.

REPUBBLICA ITALIANA (1999). Gazzetta Ufficiale 248 del 21.10.1999.

SAS INSTITUTE INC. CARY. NC USA The SAS system for Windows 1999-2001.

SBOARINA C, CESCATTI A (2004) Il clima del Trentino. Distribuzione spaziale delle principali variabili climatiche. Centro di Ecologia Alpina, report 33 and CD-ROM, Trento, I

SCATTOLIN L (2007). VARIATIONS OF THE ECTOMYCORRHIZAL COMMUNITY IN HIGH MOUNTAIN NORWAY SPRUCE STANDS AND CORRELATIONS WITH THE MAIN PEDOCLIMATIC FACTORS. PhD thesis - Università degli Studi di Padova and Ludwig-Maximilians-Universität München.www.edoc.ub.-muenchen.de/6725/1/Scattolin_Linda.pdf

SCATTOLIN L, BOLZON P, MONTECCHIO L (2008a). A geostatistical model to describe root vitality and ectomycorrhization in Norway spruce. *Plant Biosystem*, in press.

SCATTOLIN L, MONTECCHIO L, AGERER R (2008b). The ectomycorrhizal community structure in high mountain Norway spruce stands. *Trees*. DOI 10.1007/s00468-007-0164-9.

SHANNON CE, WEAVER W (1949). The mathematical theory of communication. The University of Illinois, Chicago, London. pp. 3-24.

SHI L, GUTTENBERGER M, KOTTKE I, HAMPP R (2002). The effect of drought on mycorrhizas of beech (*Fagus sylvatica* L.): changes in community structure, and the content of carbohydrates and nitrogen storage bodies of the fungi. *Mycorrhiza* (12):303–311.

SKINNER MF, BOWEN GD (1974). The penetration of soil by mycelial strands of ectomycorrhizal fungi *Soil Biology and Biochemistry* 6: 57-61.

SMITH SE, READ DJ (1997). Mycorrhizal symbiosis. San Diego, CA, USA: Academic Press.

SOKAL RR, ROHLF FJ (1981). Biometry. The principles and practice of statistics in biological research. WH Freeman and Company, San Francisco: pp. 857.

TATENO R, HISHI T, TAKEDA H (2004). Above- and belowground biomass and net primary production in a cool-temperate decidous forest in relation to topographical change in soil nitrogen. *Forest Ecology and Management* 193: 297-306.

TAYLOR AFS (2002) Fungal diversity in ectomycorrhizal communities: sampling effort and species detection. *Plant and Soil* 244: 19–28

TEDERSOO L, SUVI T, LARSSON E, KÕLJALG U (2006). Diversity and community structure of ectomycorrhizal fungi in a wooded meadow. *Mycological Research* IIO(2006): 734-748.

TER BRAK CJF (1986). Canonical correspondence analysis: a new eigenvector technique for multivariate direct gradient an*alysis*. *Ecology* 67:1167-1179.

TER BRAK CJF (1994). Canonical community ordination. Part I: Basic theory and linear methods. *Ecoscience* 1: 127-140.

TOLJANDER JF, EBERHARDT U, TOLJANDER YK, PAUL LR, TAYLOR AFS (2006). Species composition of an ectomycorrhzial fungal community along a local nutrient gradient in a boreal forest. *New Phytologist* 170: 873-884.

UNESTAM T (1991). Water repellency, mat formation, and leaf-stimulated growth of some ectomycorrhizal fungi. *Mycorrhiza*, 1: 13-20.

VAN DER HEIJDEN EW, VRIES FWDE, KUYPER THW (1999). Mycorrhizal associations of *Salix repens* L. communities in succession of dune ecosystems. I. Above-ground and below-ground views of ectomycorrhizal fungi in relation to soil chemistry. *Canadian Journal of Botany* 77:1821–1832.

WERNER A, ZADWORNY M, IDZIKOWSKA K (2002) Interaction between *Laccaria laccata* and *Trichoderma virens* in co-culture and in the rhizosphere of *Pinus sylvestris* grown in vitro. *Mycorrhiza* 12: 139-145.

WARGO PM (1988). Root vitality and mycorrhizal status of different health classed of red spruce trees. *Phytophatology* 78: 1533.

Site	Last cut (year)	Altitude (m a.s.l.)	Slope (degrees)	Exposure (degrees)	Humus	Bedrock
1	1958	1157	6	116	Dysmull	Calcareous
2	1960	1134	37	166	Amphimull	Dolomitic
3	1985	1165	26	122	Amphimull/Dysmull	Dolomitic
4	2001	1166	16	105	Dysmull	Calcareous
5	1958	1180	5	107	Amphimull/Dysmull	Calcareous
6	1982	1200	17	110	Dysmull	Calcareous
7	2004	1050	14	103	Dysmull	Calcareous

Tab. 1: **Main features of the 7 investigated stands.**

Anatomotypes and codex	Abbr.	Freq. ᴬ (2005, 2006)	June 2005		October 2005		June 2006		
			MID	Fo	MID	Fo	Freq.	MID	Fo
(EDM47)	An47	21.006	12.826	11.497	-152.25	0.940*	2.569	-17.012	0.790*
Amphinema sp.	Amphin	1.142	-	-	-141.375	0.882*	-	-	-
Boletaceae	Bol1	10.383	-	-	30.588	2.604	4.416	-11.255	0.689*
Boletus sp.	Bolrodo	0.32	109.8	9.840	-152.25	0.940*	0.5	-	-
Byssocorticium atrovirens	Byssatr	90.246	3.653	7.342	5.873	1.876	-	-	-
Cenococcum geophilum	Cenoc	498.056	1.140	2.590	1.222	1.307*	245.316	0.849	0.728*
*Cortinarius*1 sp.	Cor1	2.137	-	-	-9.176	0.933*	-	-	-
Cortinarius bolaris	Corbol	1.983	98.150	4.090	14.565	2.391	14.194	12.002	2.075
Cortinarius cinnabarinus	Corcinn	39.370	5.462	5.433	14.712	1.988	16.122	8.893	1.884
Cortinarius inochlorus	Corinoc	24.152	12.453	7.610	11.464	1.471	1	-	-
Craterellus sp.	Cratell	17.795	17.227	13.383	-	-	10.698	8.233	1.519
Entoloma sp.	Entol2	15.634	27.967	19.785	-	-	-	-	-
Fagirhiza arachnoidea	Faracnoid	-	-	-	-	-	5.444	20.538	1.643
Fagirhiza byssoporoides	Fbyssopo	6.775	-	-	1.828	1.027*	4.575	-4.225	0.861*
Fagirhiza cystidiophora	Fcystid	14.868	11.854	5.271	147.367	3.763	7.213	9.759	1.403*
Fagirhiza entolomoides	Entol1	72.676	10.963	15.826	3.987	1.654	18.471	4.030	1.392*
Fagirhiza fusca	Ffusca	13.346	14.647	3.115	-3.427	0.808*	10.825	7.262	1.455
Fagirhiza lanata	Flanata	8.318	182	6.837	14.084	1.553	-	-	-
Fagirhiza oleifera	Foleifer	62.116	3.691	6.738	13.786	1.632	4.944	15.549	1.425*
Fagirhiza pallida	Fpallida	37.613	5.022	6.110	-9.095	0.748*	6.714	26.187	1.183*
Fagirhiza setifera	Fsetif	64.709	4.498	7.153	6.540	1.620	4.875	38.808	2.085
Fagirhiza spinulosa	Fspinul	9.433	42.645	12.681	-24.333	0.748*	-	-	-
Fagirhiza stellata	Tom3	9.089	15.245	4.473	-12.484	0.906*	2.75	40.623	1.513
Fagirhiza vermiculiformis	Fvermi	6.248	23.636	3.690	-23.060	0.773*	-	-	-
Genea hispidula	Geneah	6.201	62.546	2.913	-8.131	0.786*	23.0583	0.806	0.968*
Hydnum sp.	Hydnum	470.461	43.590	1075.184	-	-	4.650	6.264	1.142*
Hygrophorus penarius	Hygro2	-	-	-	-	-	17.694	12.799	2.459
Hygrophorus sp.	Hygro1	40.666	7.287	9..326	8.478	1.336*	5.628	-6.266	0.750*
*Inocybe*1 sp.	Inoc1	11.994	18.871	4..553	14.631	1.422	-	-	-
Laccaria sp.	Lacc	22.523	9.450	6.507	21.166	1.558	10.958	20.253	2.420
Lactarius acris	Lacacris	14.879	-	-	6.257	1.452	4.375	8.491	1.187*
Lactarius rubrocinctus	Lrubroci	13.276	-	-	20.805	2.748	-	-	-
Lactarius sp.	Lacta1	2.592	-	-	-22.825	0.780*	-	-	-
Lactarius subdulcis	Lsubdul	61.827	6.657	7..354	10.156	2.739	47.480	3.789	1.960

Lactarius vellereus	Lvell	482.126	15.061	6.157	10.883	1.584	17.527	42.601	6.093
Piloderma croceum	Piloder	2.368	20.274	10.507	42.209	3.563	13.444	15.163	2.305
Ramaria1 sp.	Ram1	47.336	13.801	6..671	4.532	1.712	5.888	6.065	2.066
Ramaria aurea	Ramaur	5.652	74.224	11.154	-65.827	0.540*	-	-	-
Ramaria sp.	Ram2	-	-	-	-	-	1	-	-
Russula illota	Rusill	38.205	49.512	8.480	22.371	2.682	-	-	-
Russula mairei	Rusma	23.149	13.411	6.268	14.637	1.872	-	-	-
Sebacina2 sp.	Seba2	13.842	62.546	2.913	8.633	1.566	0.25	-	-
Sebacinaceae	Seba1	50.418	62.546	2.913	-16.925	0.862*	-	-	-
Tomentella4 sp.	Tom4	10.120	-	-	9.578	1.452	-	-	-
Tomentella1 sp.	Tom1	3.142	51.824	6.925	-	-	0.25	-	-
Tomentella2 sp.	Tom2	2.228	-	-	-13.386	0.897*	1	-	-
Thelephorales	Tomlo1	-	-	-	-	-	2.222	91.490	1.819
Tricholoma acerbum	Tricacer	22.848	9.191	4.103	7.620	1.425*	1.25	-95.2	0.821*
Tricholoma sciodes	Tricscio	93.017	18.157	16.615	2.228	1.506	12.325	13.858	2.078

Tab. 2: [A]Absolute abundance(June 2005 and October 2005). MID and test F for each sampling: *test F not significant ; p=0.05; df1= ∞ ; df2= q-1, with q as the number of samples.

Sampling	*Site 7 (14°)		Site 4 (16°)		Site 6 (17°)		Site 3 (26°)		Site 2 (37°)		Site 1 (6°)		Site 5 (5°)	
Spatial distribution	SH	HSH	SH	HSH	SH	HSH	SH	HSH	SH	HSH	SH	HSH	SH	HSH
June 2005	-	-	2.280	0.485	-	-	3.901	0.749	3.559	0.734	3.849	0.750	-	-
October 2005	-	-	3.911	0.740	-	-	3.759	0.745	4.278	0.798	3.717	0.719	-	-
June 2006	3.523	0.698	-	-	2.979	0.585	-	-	-	-	-	-	3.693	0.745
Vertical distribution	Site 7 (14°)		Site 4 (16°)		Site 6 (17°)		Site 3 (26°)		Site 2 (37°)		Site 1 (6°)		Site 5 (5°)	
October 2006	SH	HSH	SH	HSH	SH	HSH	SH	HSH	SH	HSH	SH	HSH	SH	HSH
O Horizon	-	-	2.077	2.103	-	-	2.813	0.640	2.318	0.458	1.827	0.380	2.659	0.540
A Horizon	-	-	0.411	0.437	-	-	3.138	0.660	2.127	0.484	1.600	0.364	2.373	0.540

Tab. 3: Richness, diversity and evenness of the EM community for each sampling: SH= Shannon-Weaver Index; HSH=Evenness [chronosequence of the sites with the slope measured with ° (older coppices to younger)].

	Sites	1	2	3	4	5	1	2	3	4	5		
	Soil horizons	O	O	O	O	O	A	A	A	A	A	MID	Fo
Abbrev	Anatomotypes												
An65	EDM65	0	0	0	10	0	0	0	0	7	5	21.142	6.558 *
An68	EDM68	0	0	7	0	0	0	0	0	0	0	74	74 *
Bol1	Boletaceae	71	160	73	0	15	4	82	15	0	0	11.273	2.445 *
Bolrodo	Boletus sp.	0	0	0	105	0	0	0	0	55	1	39.901	21.730 *
Byssatr	Byssocorticium atrovirens	23	10	9	1	1	0	1	3	9	5	10.273	2.178 *
Cenoc	Cenococcum geophilum	3330	2193	2019	3034	943	1638	1223	742	2714	445	3.778	1.105
Cor1	Cortinarius1sp.	0	0	34	21	0	0	0	10	17	0	14.884	3.640 *
Corbol	Cortinarius bolaris	34	0	16	2	0	10	37	60	10	0	6.034	1.347 *
Corcinn	Cortinarius cinnabarinus	36	32	18	0	233	15	43	173	128	113	4.306	1.149
Corinfr	Cortinarius infractus	0	28	1	8	0	0	0	1	89	0	17.976	4.947 *
Corinoc	Cortinarius inochlorus	2	15	48	22	21	3	0	10	17	0	10.081	2.129 *
Faracnoid	Fagirhiza arachnoidea	0	0	23	0	4	0	0	43	5	0	11.116	2.401 *
Fbyssop	Fagirhiza byssoporoides	0	0	18	37	0	0	0	12	1	18	12.034	2.668 *
Entol1	Fagirhiza entolomoides	0	1	73	199	55	0	56	2	72	10	11.611	2.542 *
Ffusca	Fagirhiza fusca	8	0	0	52	0	3	0	0	14	8	25.033	8.912 *
Flanata	Fagirhiza lanata	3	3	8	3	4	2	1	20	37	4	11.704	2.569 *
Foleifer	Fagirhiza oleifera	0	36	2	23	0	26	0	0	65	3	12.115	2.692 *
Fpallida	Fagirhiza pallida	106	37	6	14	5	0	0	12	16	0	14.124	3.359 *
Fsetif	Fagirhiza setifera	19	0	55	87	0	0	9	71	2	0	9.603	2.014 *
Fspinul	Fagirhiza spinulosa	12	31	0	55	0	16	0	0	41	33	13.665	3.197 *
Tom21	Fagirhiza stellata	0	80	0	0	7	0	20	66	0	7	11.119	2.402 *
Geneah	Genea hyspidula	8	0	0	18	0	4	6	0	0	0	8.164	1.703 *
Hygro1	Hygrophorus sp.	86	6	282	66	2	15	12	40	105	7	8.509	1.772 *
Lacc	Laccaria sp.	18	11	0	131	7	0	0	14	106	3	15.920	4.049 *
Lacris	Lactarius acris	140	141	15	0	31	60	44	50	0	13	9.404	1.967 *
Lpallid	Lactarius pallidus	29	101	49	175	100	0	30	26	153	32	4.389	1.157
Lsubdul	Lactarius subdulcis	50	29	118	76	50	0	58	24	112	2	5.883	1.326
Pezi1	Pezizales	10	13	0	12	0	13	26	0	0	0	10.321	2.190 *
Ramaurea	Ramaria aurea	10	0	0	0	0	0	0	0	0	0	74	74 *
Ram1	Ramaria sp.	17	4	6	7	87	15	78	19	15	53	7.700	1.615 *
Ram2	Ramaria sp.	262	0	0	8	0	42	37	1	0	0	17.049	4.528 *
Rusma	Russula mairei	6	25	1	6	3	2	6	14	10	3	19.588	5.733 *
Seba2	Sebacina2 sp.	21	33	118	45	13	23	4	84	75	0	5.166	1.237
Seba1	Sebacinaceae1	21	0	0	0	5	0	0	0	0	0	50.092	34.014 *

Teleph1	*Thelephoraceae*	0	0	0	0	66	0	0	21	0	0	30.801	13.166	*
Teleph2	*Thelephoraceae*	20	20	0	0	0	0	0	0	0	0	19.448	5.662	*
Tom1	*Tomentella1* sp.	44	15	0	0	48	5	21	0	5	3	11.014	2.373	*
Tom2	*Tomentella2* sp.	4	0	26	35	11	0	0	5	0	0	20.104	5.999	*
Tom5	*Tomentella5* sp.	0	2	0	0	0	0	0	0	0	0	74	74	*
Tricacer	*Tricholoma acerbum*	0	5	2	8	0	23	0	0	5	9	15.578	3.911	*
Tomlo1	*Thelephorales1*	0	0	0	4	0	0	0	0	0	0	74	74	*
Tomlo2	*Thelephorales2*	48	102	0	8	100	318	0	17	0	0	19.136	5.506	*
Tricscio	*Tricholoma sciodes*	27	0	0	27	0	13	9	25	8	0	7.128	1.514	*

Tab. 4: Absolute abundance of the different anatomotypes in the soil horizons, MID and test F: *test F significant ; p=0.05; df1= ∞ ; df2= q-1, with q as the number of samples.

Soil horizons		O		A	
Trees	Sites	Vital	Non-vital	Vital	Non-vital
a	1	3294	1799	1540	629
b	1	1171	665	710	289
a	2	2254	1003	607	303
b	2	1484	1918	34	40
a	3	1271	2927	997	1737
b	3	1756	537	583	739
a	4	1832	1023	2150	1372
b	4	2496	1613	1743	701
a	5	1038	692	473	554
b	5	773	998	304	378

Tab. 5: Absolute abundance of the vital and non-vital EM in the vertical distribution in the soil.

Fungal taxa	Best match sequence	1c	2c	3c	Sitze (pair)	E value	Similarity	Accession number	Source(a)
Amphinema sp.	-		x		-	-	-	-	-
Boletaceae1	*Boletus aestivalis*	x	x	x	**	3E-73	90%	EU444544	UNITE UDB000941
Boletus2 sp.	*Boletus rhodoxanthus*	x	x	x	661	0.0	99%	EU444539	UNITE UDB001116
Byssocorticium atrovirens	-		x	x	x	-	-	-	-	-
Cenococcum geophilum	-		x	x	x	-	-	-	-	-

Cortinarius inochlorus	Cortinarius ionochlorus	x	x	x	601	0.0	100%	EU444542	UNITE UDB002105
Cortinarius1 sp.	-	x	x		-	-	-	-	-
Cortinarius bolaris	-	x	x	x	-	-	-	-	-
Cortinarius cinnabarinus	-	x	x	x	-	-	-	-	-
Craterellus sp.	-	x	x		-	-	-	-	-
Cortinarius infractus	Cortinarius infractus			x	541	0.0	100%	EU444553	UNITE UDB001161
EDM47	-	x	x		-	-	-	-	-
EDM65	-	-	-	-	-	-	-	-	-
EDM68	-	-	-	-	-	-	-	-	-
Entoloma2 sp.	-	x			-	-	-	-	-
Entolomataceae1 (Fagirhiza entolomoides)*	Entoloma sp.	x	x	x	901	e-168	91%	EU444549	UNITE UDB000937
Fagirhiza arachnoidea	-	x	x		-	-	-	-	-
Fagirhiza byssoporoides *	Byssoporia terrestris fruitbody (SR1101 in M)	x	x		541	-	99%	EU444550	-
Fagirhiza cystidiophora	-	x	x	x	-	-	-	-	-
Fagirhiza fusca	-	x	x	x	-	-	-	-	-
Fagirhiza lanata	-	x		x	-	-	-	-	-
Fagirhiza oleifera	-	x	x	x	-	-	-	-	-
Fagirhiza pallida	-	x	x	x	-	-	-	-	-
Fagirhiza setifera	-	x	x		-	-	-	-	-
Fagirhiza spinulosa	-	x		x	-	-	-	-	-
Tomentella3 (Fagirhiza stellata) *	Tomentella subtestacea	x	x	x	661	0.0	92%	EU444548	UNITE UDB000034
Fagirhiza vermiculiformis	-	x			-	-	-	-	-
Genea hispidula	-	x	x	x	-	-	-	-	-
Hydnum sp.	-	x	x		-	-	-	-	-
Hygrophorus1 sp.	Hygrophorus sp.	x	x	x	**	8e-75	96%		UNITE UDB000556
Hygrophorus penarius *	Hygrophorus penarius		x		481	0.0	100%	EU444536	UNITE UDB000097
Inocybe1 sp.	-	x			-	-	-	-	-
Laccaria sp.	-	x	x	x	-	-	-	-	-
Lactarius acris	-	x	x	x	-	-	-	-	-
Lactarius pallidus	-	x	x	x	-	-	-	-	-
Lactarius rubrocinctus	-	x			-	-	-	-	-
Lactarius1 sp.	-	x			-	-	-	-	-

Lactarius subdulcis	-	x	x	x	-	-	-	-	-
Lactarius vellereus	-	x	x		-	-	-	-	-
Piloderma croceum	-	x	x		-	-	-	-	-
Ramaria aurea	-	x		x	-	-	-	-	-
Pezizales1	Peziza sp.				**	3e-57	91%	EU444547	UNITE UDB001572
Ramaria2 sp.	-	x	x		-	-	-	-	-
Ramaria1 sp.	Albatrellus critstatus	x	x	x	331	2e-91	98%	EU444537	UNITE UDB001761
Russula illota	-	x			-	-	-	-	-
Russula mairei	-	x	x	x	-	-	-	-	-
Sebacina2 sp.	Uncultured ectomycorrhiza (Sebacinaceae)	x	x	x	541	0.0	95%	EU444543	BLAST AJ879661
Sebacinaceae1	Sebacina epigea	x		x	541	0.0	94%	EU444538	UNITE UDB000975
Thelephoraceae1	-	-	-	-	-	-	-	-	-
Thelephoraceae2	-	-	-	-	-	-	-	-	-
Thelephorales1	-		x		**	-	-	EU444545	-
Thelephorales2	Tomentellopsis echinospora			x	541	0.0	94%	EU444546	UNITE UDB000191
Tomentella1 sp.	Tomentella cinerascens	x	x	x	481	0.0	99%	EU444540	UNITE UDB000232
Tomentella2 sp.	Tomentella pilosa	x	x	x	601	0.0	97%	EU444541	UNITE UDB000241
Tomentella4 sp.	-	x			-	-	-	-	-
Tomentella5 sp.	-	x	x		-	-	-	-	-
Tricholoma acerbum	-	x	x		-	-	-	-	-
Tricholoma sciodes	-	x	x	x	-	-	-	-	-

Table 6: EM anatomotypes: anatomical, morphological and molecular identification. (a) Reference available on NCBI (www.ncbi.nih.gov/BLAST) or UNITE (www.unite.ut.ee) websites [* Description in progress;** Partial sequence available; 1c = first collection (2005); 2c = second collection (2006); 3c = third collection (vertical distribution); x = EM presence].

CCA Spatial distribution 2005

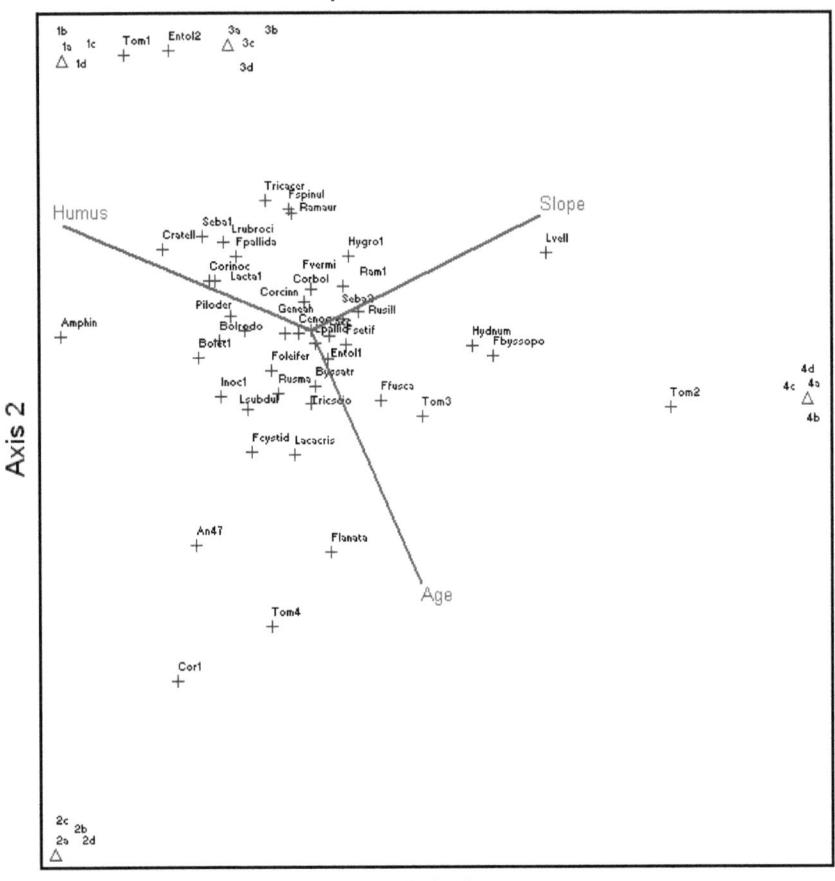

Fig. 1: CCA joint biplot of the EM fungal community in the 1, 2, 3, 4 stands. Open triangles represent sampling position [a, b, c, d= stump; for EM species abbreviation see tabs. 1, 3]. Vectors indicate quantitative parameters [Slope; Age = coppice age; Humus = hymus types; for example: 2b = site 2, stump b]. Correlation measured ["intraset correlations" (ter Braak, 1986)]: to slope with a value of 0.650, to the age with a value of 0.312; negative correlations mainly with humus form with a coefficient of -0.706.

Variable	Axis 1	Axis 2	Axis 3
1. Age	0.312	- 0.762	0.343
2. Humus	- 0.706	0.318	0.322
3. Slope	0.650	0.342	0.431

Tab. 7: Intraset correlations (Ter Braak 1986) Spatial distribution 2005.

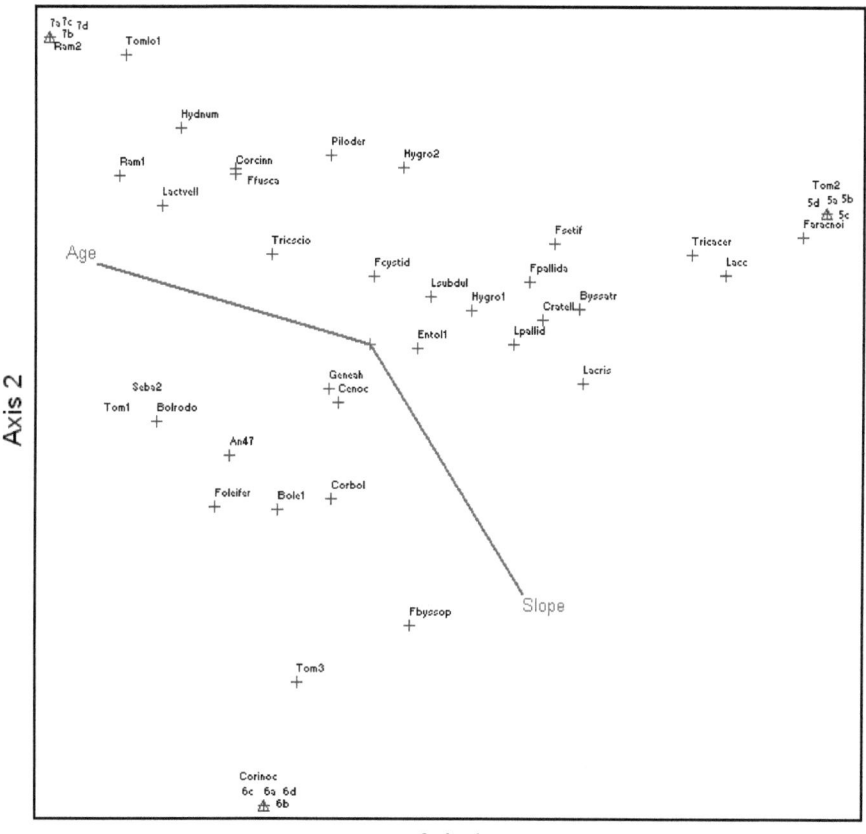

Fig. 2: CCA joint biplot derived from the EM fungal community in the 5, 6, 7 stands. Open triangles represent the sampling position [a, b, c, d = stump; for abbreviation see tabs, 1, 3]. Vectors indicate quantitative parameters (Slope; Age: coppice age; see Table 1).

Variable	Axis 1	Axis 2	Axis 3
1. Age	-0.890	0.239	0.000
2. Slope	-0.503	-0.747	0.000

Tab. 8. Intraset correlations (Ter Braak 1986) Spatial distribution 2006.

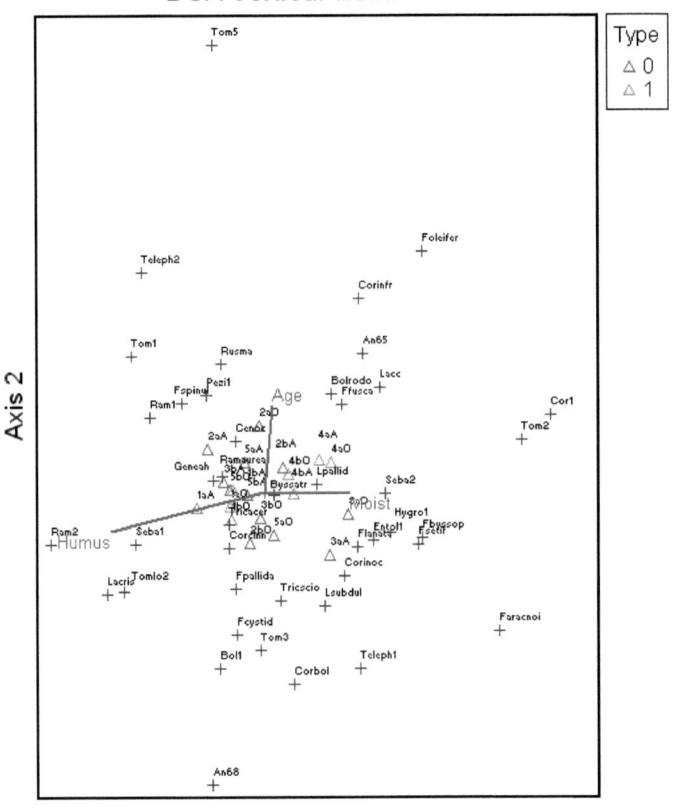

Fig. 3: DCA joint biplot of the EM fungal community, assembling in the data-set the sampling directions and the soil horizons in the 1, 2, 3, 4, 5 sites. Open triangles with different colours represent the sites with different bedrock types (for the site feature see tab. 1; for the species abbreviations see tab. 3). The vectors indicate the direction of the gradient explained with the following ecological factors: Moist = sample soil moisture; Age = coppice age; Humus = humus types; for example: 4aA = site 4, stump a, soil horizon A].

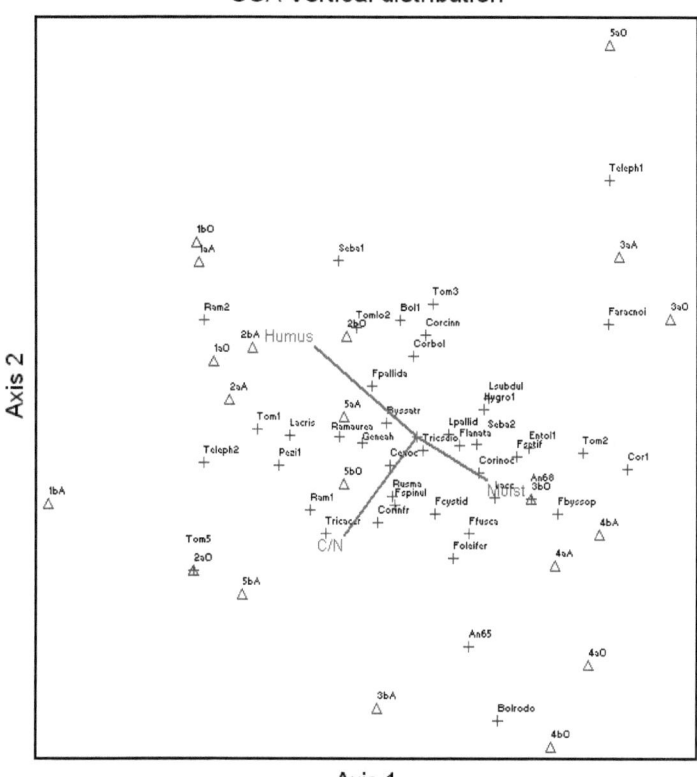

Fig. 4: CCA joint biplot derived from the EM fungal community in the 1, 2, 3, 4, 5 coppice stands. Open triangles represent sampling position [a, b = stump; for abbreviation see tabs. 1, 3]. Vectors indicate quantitative parameters [Moist = sample soil moisture; C/N = ratio C_{org} / N_{tot}; Humus = humus types; for example: 4aA = site 4, stump a, soil horizon A]. Correlation measured ["intraset correlations" (ter Braak, 1986)]: to soil moisture with a value of 0.470, to the slope with a value of 0.290, and to the pH with a coefficient of 0.166; negative correlations mainly with humus form with a coefficient of -0.684, with C/N ratio (- 0.482), with altitude (-0.110) and age (shoot age, -0.087).

Variable	Axis 1	Axis 2	Axis 3
1.Ae	- 0.083	-0.440	0.194
2. Moist	0.447	-0.262	0.207
3. C/N	-0.458	-0.604	0.446
4. pH	0.158	0.228	0.110
5. Humus	-0.651	0.551	0.226
6. Alt	-0.104	-0.064	0.779
7. Slope	0.276	-0.415	0.401

Tab. 9. Intraset correlations (Ter Braak 1986) Vertical distribution 2006.

Short communication

Distribution of ectomycorrhizae in Beech coppices regarding exploration types and hydropobicity-hydrophily features

Erika Di Marino[1,2,3], Lucio Montecchio[1], Linda Scattolin[2], Clemens Abs[4], Reinhard Agerer[3]

[1] *Università degli Studi di Padova, Dipartimento Territorio e Sistemi Agroforestali, V.le dell'Università, 16, I-35020 Legnaro (PD), Italy.*

[2] *Centro di Ecologia Alpina, Viote del Monte Bondone, I- 38070 Sopramonte (TN), Italy.*

[3] *Organismic Biology: Mycology, Department Biology and GeoBio-Center[LMU], University of München, Menzinger Str., 67, D-80638, Germany*

[4] *Fachgebiet Geobotanik, Wissenschaftszentrum Weihenstephan, Technische Universität München,, Am Hochanger 13, D-85354 Freising, Germany*

A rapid increase of scientific interest in the development and application of ecological indicators in the last 40 years, contributed to focus the researches on the measurement of indicator species to show similar ecological requirements. Detailed studies also of ectomycorrhizae (EM) of forest sites were initiated in the 1990s, in order to extend the results to bioindication. Here we report the preliminary investigations on ecological features of EM species in seven Beech coppices of North Italy with different turns. Multivariate analyses (DCA and CCA) were performed to investigate the spatial distribution and the vertical distribution to test the correlation of the species distribution with the exploration attitude and the hydrophilic or hydrophobic behaviour. The results attested a prevailing presence of hydrophilic species and a probably attitude to use the "medium-distance" exploration strategy. Interpretations of these correlations are still difficult. Further, similar studies in combination with the analysis of soil factors will possibly unravel the complex situation. No clear correlation was observed of the putatively ecologically important EM features with the age of the coppices.

Introduction

The past 40 years have seen a rapid increase of scientific interest in the development and application of ecological indicators. This focus on indicators derives from the need to assess ecological conditions to make regulatory, stewardship, sustainability, or biodiversity decisions. (Niemi et al., 2004). Environmental indicators should reflect all the elements of the causal chain that link human activities to their ultimate environmental impacts, and the societal responses to these impacts (Smeets & Weterings, 1999). Ecosystem disturbance can be natural (e.g., fire, wind, and drought) and part of the functional attributes of ecosystems (Noss 1999), or it can be anthropogenic like repeated coppicing.

Most applications of ecological indicators have focused at the species level due to concerns arising from endangered species and species conservation issues (Fleischman et al., 2001). The measurement of an indicator species assumes that a single species represents many species with similar ecological requirements (Landres et al., 1988). As a consequence, detailed studies of ectomycorrhizae (EM) of forest sites were initiated in the 1990s, in order to apply the results to bioindication (Al Sayegh Petkovšek 1997; 2004; 2005; Al Sayegh Petkovšek & Kraigher, 2003; Erland & Taylor, 2002; Fellner & Peškova, 1995; Kraigher et al., 1996; Kraigher et al., 2006; Taylor & Alexander, 2005; Tayloret al., 2000). Many EM features are functionally important and they seem to play a particular ecological role: EM mantles could be a shelter against microbial attack (Werner et al., 2002) and might be a buffer against rapid loss of water, when the mantle hyphae form a gelatinous matrix (Agerer 2006), or they may provide a suitable surface for bacteria (Mogge et al., 2000; Schelkle et al., 1996; Timonen et al., 1998) that might be helpful for the formation of EM (Garbaye & Dopunnois, 1993) or for fixation of nitrogen (Amaranthus et al. 1990). Smooth and hydrophilic mantles can directly acquire water and nutrients, while hydrophobic EM with well developed rhizomorphs can transport nutrients over distances of several decimetres (Kammerbauer et al., 1989; Schramm 1966; Skinner & Bowen, 1974, Unestam 1991; Agerer 2001). The species, which possess water repellence properties, seem to prefer highly areated soil in the conifer forest soils (Unestam 1991). In contrast to this behaviour, the ecological strategy of the hydrophilic EM is not very clear (Unestam 1991; Unestam & Stenström,1989; Stenström 1991). These hydrophilic mantles (e.g. many *Lactarius* species) appear to be a close control over the movement and the exchange of material through the mantle (Ashford et al., 1988), and are most likely responsible for the uptake of water and nutrients (Cairney & Burke, 1996).

EM fungi probably control the interface between the soil environment and the host plant: the mantles may control the fluxes into and out of the root, the mycelium extending out from the mantle surface in the surrounding soil (the extramatrical mycelium) is considered to be the primary site for nutrient and water uptake (Taylor & Alexander, 2005).

Some researches revealed that ectomycorrhizal species differ in their ability to exploit soil nutrients developing a range of anatomical structures (Agerer 2001) and this diversity might explain their distribution among different ecological niches (Bruns 1995; Dickie *et al.*, 2002; Erland & Taylor, 2002; Agerer 2006).

The extension and the structure of this extramatrical mycelium is thought to be different among EM fungal taxa (Agerer 2001; 2006). In this context the proposal to classify the EM fungal species according to their "exploration types" (Agerer 2001) by interpreting the anatomical features as ecological strategies to colonise the soil, becomes more and more important to understand the role of these organisms, appearing as key elements of forest nutrient cycles and strong diversity of forest ecosystem processes (Read *et al.*, 2004).

The mycelium formed hydrophilic structures seems to have substrate particles glued to their surface (Raidl 1997). Here we report on preliminary investigations on the ecological features of EM species (hydrophobicity and exploration types) in Beech coppices of North Italy.

Experimental design

The investigations on the spatial distribution were performed in 2005 and 2006 in 7 coeval beech [*Fagus sylvatica* L.] coppices 2- to 48-years-old growing in the Natural Park of Adamello-Brenta (Northern Italy; 5.125.228 ÷ 5.125.666 N, 1.654.361 ÷ 1.654.565 E), selected among the most productive and exploited in the Trentino-Südtirol Region (beech presence 85-90% Provincia Autonoma di Trento, 2001; Sboarina & Cescatti, 2004, climatic conditions tab.1).

From these, in 2005, 4 sites differing in age of coppicing (coppiced in 1958, 1952, 1980, 2001, respectively) and bedrock type (dolomitic, calcareous) were selected and coded 1 to 4. In 2006, in order to verify the extendibility of the obtained results, 3 additional comparable stands growing at least 5 km far from the firsts were selected and coded 5 to 7 (coppiced in 1958, 1982, 2004, respectively).

In each plot, 4 stumps apparently healthy, undamaged by climatic events, at least 15 m from the nearest EM tree, were randomly selected and coded. In June and October 2005 (sites 1÷4) and in June 2006 (sites 5÷7), from each stump 12 cylindrical soil cores (18 mm diameter; 15 cm deep) were collected (100, 150 and 200 cm from the collar, along N, E, S and W directions) and stored in plastic pipes at 4 ±1 °C in the dark. For each core, the humus form was classified according to Jabiol *et al.* (1995).

In October 2006, investigation were performed on the EM vertical distribution in the sites 1÷5. Soil samples of 2.5 x 2.5 cm were collected up to lowest mineral layer A (including the litter layer). The samples were collected at 150 cm from the base (below the canopy projection) and along the four cardinal directions. The organic horizon O and the mineral horizon A were accurately classified, and each sample was preserved as reported above.

To investigate the spatial distribution, within 12 days from sampling, 10 rootlets with undamaged and fully developed apical tips were randomly chosen from every soil core and carefully cleaned. For each rootlet the last apex was distinguished as *not vital* (NV), *vital not-mycorrhizal* (NM), and *vital ectomycorrhizal* (EM). For the present analyses, only the vital ectomycorrhizal tips were considered.

Every EM apex was classified anatomically and morphologically (Goodman *et al.*, 1996; Agerer 1987-2006; Brand 1991; Agerer & Rambold, 2004-2007), and the ones with uncertain classification were submitted to molecular analyses (Gardes & Bruns, 1993; Beenken 2004; Tedersoo *et al.*, 2006), using 10 mycorrhizal apexes per anatomotype. The hydrophobicity according to Unestam (1991)and Agerer (2006) and the exploration types according to Agerer (2001) were also checked.

DNA extraction, amplification, sequencing and assignation of sequence to taxa were performed according to Mosca *et al.* (2007). Anatomotypes detected with not enough apexes to allow the molecular procedure after the morphological one, and the ones which ITS sequence gave uncertain results were classified by an alphanumerical code (EDMxx).

All specimens were preserved in FEA (formaldehyde 40% : ethyl alcohol 50% : acetic acid 100% : = 5 : 90 : 5, v/v/v) solution and stored in the TeSAF Departmental herbarium, University of Padova.

Detrended Correspondence Analysis (DCA; Hill and Gauch, 1980) and Canonical Correspondence Analysis (CCA; Hill, 1979) were carried out considering the *absolute abundance* of EM in each sample (comp. Scattolin *et al.*, 2008, total number of tips/soil core). A power transformation (power = 0.50, square root) was applied to the data set only for DCA reducing the number of the interactions and then applied also to perform the CCA. The 2 types of analyses were performed using PC-ORD^{TM} (McCune & Mefford, 1999, version 5 for Windows, MjM, Oregon).

For the vertical distribution the *relative abundance* (Σ EM/cm^3 soil volume and Σ NVM/cm^3 soil volume) were calculated and used in the multivariate analyses. Due to the structure of the data-set, data regarding the sampling directions were gathered, as no significant differences were found in the EM community for this parameter.

Relations between the ecological features (the hydrophobicity, according to Unestam 1991 and the exploration types according to Agerer 2001) and the species distribution, were tested using the Detrended Correspondence Analysis (DCA; Hill & Gauch, 1980) and the Canonical Correspondence Analysis (CCA; Hill, 1979), applying a power transformation (power=0.50, square root). To quantify the ecological factors we used these values: species hydrophobic = 0; species hydrophilic = 1; exploration type = 1: C = contact type; 2: SD= short distance; 3 : MD sm= medium distance smooth; 4 : MD fr= medium distance fringe; 5 : MD mat= medium distance mat; 6 : LD= long distance; 7: C/SD = between C/SD. To understand the correlation between the factors and the species, it is important to follow the vector direction: the values of the vectors grow with the distance from the centroid (i.e. at the end of the vectors it can be found the hydrophilic species (value "1" of the first vector) with a C/SD strategy (value "7" of the second vector).

Due to the structure of the data-set, obtained from the investigation in the soils (in the year 2006, on the 1, 2, 3, 4, 5 sites) which consisted of several plots and rare species with low abundances and only with one dominant species, we assembled the data regarding the sampling directions.

Results and discussion

The DCA performed on the data-set collected in the year 2005 and 2006 demonstrated no gradients in the EM spatial distribution related to the exploration types and the hydrophobicity (data not shown).

But in the CCA results (fig.1) concerning the first collection (June 2005) in the sites S1, S2, S3, S4, the species were correlated to the exploration types, while the hydrophilic attitude gave lower effects on the EM species distribution (total inertia = 0.4377 ; eigenvalue of the 1st and 2nd axis 0.058 and 0.002, respectively). The correlation measured on the first axis ["intraset correlations" (ter Braak, 1986)] showed that the species distribution was highly correlated with the exploration types (0.360), and a low negative correlation was revealed with the hydrophilic attitude (-0.430). The species on the right part of the graphic showed a hydrophilic attitude and mainly a "medium-distance" exploration type, but it was difficult to find a real correlation with the site features, because they characterized only lowly the calcareous sites (S2, S3 with last cut in the year 1958 and in the year 2001 respectively).

Contrasting results were obtained using the data-set of the second collection (October 2005, in the same sites). In the CCA (fig. 2) it is clear that the two ecological features investigated, were not well correlated with the species distribution. The total inertia measured was 0.3276; the eigenvalue of the 1st and 2nd axis 0.024 and 0.002, respectively. The species distribution is the same of the preceding summer: the hydrophobic species were separated from the hydrophilic species, but there are no positive correlations with the two mycorrhizal features investigated (- 0.335 for the hydrophobicity and – 0.069 for the exploration type).

In the summer of the year 2006 the situation is very similar to that revealed in the preceding summer in the sites S5, S6, S7. The CCA (fig.3) showed a strong positive correlation with hydrophobicity behaviour. The total inertia measured was 0.2229 and the eigenvalue of the 1st and

2nd axis 0.007 and 0.002, respectively. The correlation coefficients were very low: for the hydrophobicity it was 0.246, while for the exploration types -0.190. No clear correlation of the EM species and the site conditions had been obtained.

A stronger correlation with the hydrophobicity was obtained with the vertical distribution: the total inertia was 0.5571, while the correlation with the hydrophobicity was 0.619 and -0.428 with the exploration types. The eigenvalue of the 1st and 2nd axis were 0.057 and 0.010, respectively.

The EM species distribution seemed to be independent from the shoots age and from the environmental conditions measured up to now (results not shown), but more frequently correlated to the hydrophilic attitude (figs. 1, 3, 4), although the precipitation decrease was high in the two sampling years (tab. 1). Only in October 2005 the exploration types confirmed a better correlation but not very significant with the EM species (fig. 2). Although the correlations with the sites were never high, the EM species with hydrophilic attitude and of the "medium-distance" strategy seemed to prefer the site S4 and S1 (dolomitic sites). The EM species formed for these reasons "micro-communities", that remained always constant. In the different collection times changed only the rate of the correlation with the ectomycorrhizal features. Further investigations are also necessary to understand whether or not there is a correlation of hydrophobicity/hydrophily and exploration types with the vertical distribution of EM species.

In conclusion, these preliminary results could only attest a prevailing presence of hydrophilic species and a probably attitude to use the "medium-distance" exploration strategy, in the soil of beech coppices. Interpretations of these correlations are still difficult. Further, similar studies in combination with the analysis of soil factors will possibly unravel the complex situation. No clear correlation was observed of the putatively ecologically important EM features with the age of the coppices.

References

AGERER R (ED) (1987-2006). Colour atlas of ectomycorrhizae. 1-12th delivery, Einhorn, Schwäbisch Gmünd, D

AGERER R (2001). Exploration types of ectomycorrhizae: a proposal to classify ectomycorrhizal mycelial systems according to their patterns of differentiation and putative ecological importance. *Mycorrhiza* 11: 107-114.

AGERER R (2006). Fungal relationship and structural patterns of their ectomycorrhizae. *Mycological Progress* 5(2): 67-107.

AGERER R and RAMBOLD G (2004–2007) [update 2004-06-05] DEEMY – An Information System for Characterization and Determination of Ectomycorrhizae. www.deemy.de - München, D

AL SAYEGH PETKOVŠEK S (1997). Mycorrhizal potential of two differently polluted forest sites in the emission region of the Thermal Power Plant Šoštanj. *Zbornik gozdarstva in lesarstva* 52: 323–350.

AL SAYEGH PETKOVŠEK S (2004). Biodiversity of types of ectomycorrhizae in fagus stands in differently polluted forest research plots. *Zbornik gozdarstva in lesarstva* 75: 5–19.

AL SAYEGH PETKOVŠEK S (2005). Belowground ectomycorrhizal fungal communities at fagus stands in differently polluted forest research plots. *Zbornik gozdarstva in lesarstva* 76: 5–38.

AL SAYEGH PETKOVŠEK S, KRAIGHER H (2003). Mycorrhizal potential of two forest research plots with respect to reduction of the emissions from the Thermal Power Plant Šoštanj. *Acta Biologica Slovenica* 46: 9–16.

AMARANTHUS M P, LI CY, PERRY DA (1990). Influence of vegetation type and madrone soil inoculum on associative nitrogen fixation in Douglas-fir rhizosphere. *Canadian Journal of Forest Research* 20: 368-371.

ASHFORD AE, PETERSON CA, CARPENTER JL, CAIRNEY JWG, ALLAWAY, WG (1988). Structure and permeability of the fungal sheath in the *Pisonia mycorrhiza*. *Protoplasma* 147: 149-161.

BENKEEN L (2004). Die Gattung *Russula*. Untersuchungen zi ihrer Systematik anhand von Ektomykorrizen. Dissertation zur Erlangung des Grades eines Doktors der Naturwissenschaften der Fakultät Biologie der Ludwig-Maximilians-Universität München.

BRAND F (1991). Ektomykorrhizen an *Fagus sylvatica*. Charakterisierung und Identifizierung, ökologische Kennzeichnung und unsterile Kultivierung. Libri Botanici, IHW-Verlag : 1-229.

BRUNS TD (1995). Thoughts on the processes that maintain local species diversity of ectomycorrhizal fungi. *Plant and Soil* 170: 63-70.

CAIRNEY JWG, BURKE RM (1996). Physiological hetereogeneity within fungal mycelia: an important concept for a functional understanding of the ectomycorrhizal symbiosis. *New Phytologist* 134: 685-695.

DICKIE IA, XU B, KOIDE RT (2002). Vertical niche differentiation of ectomycorrhizal hyphae in soil as shown by T-RFLP analysis. *New Phytologist* 156: 527-535.

ERLAND S, TAYLOR AFS (2002). Diversity of Ectomycorrhizal Fungal Communites in Relation to the Abiotic Environment. In: *Ecological Studies* Vol. 157 M.G.A. Van der Heijden, I Sanders (Eds.) Mycorrhizal Ecology . Springer-Verlag Berlin Heidelberg.

FLEISCHMAN E, BLAIR RB, MURPHY DD (2001). Empirical validation of a method for umbrella species selection. *Ecological Applications* 11:1489–501.

FELLNER R, PEŠKOVA V (1995). Effects of industrial pollutants on ectomycorrhizal relationships in temperate forest. *Canadian Journal of Botany* 73 (Suppl. 1): 1310-1315.

GARBAYE J, DUPONNOIS R (1993). Specificity and function of mycorrhization helper bacteria (MHB) associated with *Pseudotsuga menziesii Laccaria laccata* symbiosis. *Symbiosis* 14(1-3): 335-344.

GARDES M AND BRUNS TD (1993). ITS primers with enhanced specificity for basidiomycetes – application to the identification of mycorrhizae and rusts. *Molecular Ecology* 2: 113–118.

GOODMAN DM, DURALL DM, TROFYMOW JA, BERCH SM (eds) (1996). A manual of concise descriptions of north american ectomycorrhizae: including microscopic and molecular characterization. *Mycologue Publications and the Canada-BC Forest Resource Development Agreement, Pacific Forestry Centre,* Victoria, B.C

HILL MO (1979). DECORAN-A Fortran program for detrended corrispondence analysis and reciprocal averaging. Ecology and Systematics, Cornell University, Ithaca New York.

HILL MO, GAUCH HG 1980. Detrended correspondence analysis: an improved ordination technique. *Vegetatio* 42: 47-58.

ISTITUTO AGRARIO SAN MICHELE ALL'ADIGE (Provincia di Trento) Italy - www.iasma.it

JABIOL B, BRETHES A, PONGE JF, TOUTAIN F, BRUN JJ (eds) (1995). L'humus sous toutes ses formes. ENGREF, Nancy, F

KAMMERBAUER H, AGERER R, SANDERMANN H (1989). Studies on ectomycorrhiza XXII. Mycorrhizal rhizomorphs of *Telephora terrestris* and *Pisolithus tinctorius* in association with Norway spruce *(Picea abies):* formation in vitro and translocation of phosphate *Trees*: 78-84.

KRAIGHER H, AL SAYEGH PETKOVŠEK S, GREBENC T, SIMONČIČ P (2006). Types of Ectomycorrhiza as Pollution Stress Indicators:Case Studies in Slovenia. *Environ Monit Assess* DOI 10.1007/s10661-006-9413-4.

KRAIGHER, H, BATIČ F, AGERER R (1996). Types of ectomycorrhizae and mycobioindication of forest site pollution. *Phyton (Horn, Austria)* 36: 115–120.

LANDRES PB, VERNER J, THOMAS JW(1988). Ecological uses of vertebrate indicator species: a critique. *Conservation Biology* 2:1–13.

McCUNE B, MEFFORD MJ (1999). PC-ORD^{TM}. Multivariate Analysis of Ecological Data. Version 5 for Windows, MjM Software Gleneden Beach Oregon.

MOGGE B, LOFERER C, AGERER R, HUTZLER P (2000). Bacterial community structure and colonization patterns of *Fagus sylvatica* L. ectomycorrhizospheres as determined by fluorescence in situ hybridization and confocal laser scanning microscopy. *Mycorrhiza* 9(5): 271-278.

MOSCA E, MONTECCHIO L, SELLA L, GARBAYE J (2007a).Short-term effect of removing tree competition on the ectomycorrhizal status of a declining pedunculate oak forest (*Quercus robur* L.). *Forest Ecology and Management* 244: 129-140.

NIEMI JG, McDONALD EM (2004).Application of ecological indicators. *Annual Review of Ecology Evolution and Systematics* 35:89–111.

NOSS RF. 1999. Assessing and monitoring forest biodiversity: a suggested framework and indicators. *Forest Ecological Management* 115:135–46.

PROVINCIA AUTONOMA DI TRENTO (2001) I dati della pianificazione forestale aggiornati al 31/12/2000. Servizio Foreste-Sistema Informativo Ambiente e Territorio. CD-ROM, Trento, I

RAIDL S (1997). Studien zur Ontogenie an Rhizomorphen von Ektomykorrhizen. Brbl Mycol 169:1–184.

READ DJ, LEAKE JR, PEREZ-MORENO J (2004). Mycorrhizal fungi as drivers of ecosystem processes in heathland and boreal forest biomes. *Canadian Journal of Botany* 82: 1243-1263.

SBOARINA C, CESCATTI A (2004) Il clima del Trentino. Distribuzione spaziale delle principali variabili climatiche. Centro di Ecologia Alpina, report 33 and CD-ROM, Trento, I

SCATTOLIN L, MONTECCHIO L, AGERER (2008). The ectomycorrhizal community structure in high mountain Norway spruce stands. Trees. DOI 10.1007/s00468-007-0164-9.

SCHELKLE M, URSIC M, FARQUHAR M, PETERSON RL (1996).The use of laser scanning confocal microscopy to characterize mycorrhizas of *Pinus strobus* L. and to localize associated bacteria. *Mycorrhiza* 6: 431-440.

SCHRAMM JR (1966). Plant Colonization Studies on Black Wastes from Anthracite Mining in Pennsylvania *Transactions of the American Philosophical Society*, New Ser., 56 (1) pp. 1-194.

SKINNER MF, BOWEN GD (1974). The penetration of soil by mycelial strands of ectomycorrhizal fungi *Soil Biology and Biochemistry* (6): 57-61.

SMEETS E, WETERINGS R(1999). Environmental indicators: typology and overview. *Tech. Rep. 25*, Eur. Environ. Agency, Copenhagen, Den. http://reports.eea.eu.int:80/TEC25/en/ tech 25 text.pdf.

STENSTRÖM E (1991) The effects of flooding on the formation of ectomycorrhizae in *Pinus sylvestris* seedlings. *Plant Soil* 131:247-250.

TAYLOR AFS, ALEXANDER IJ (2005). The ectomycorrhizal symbiosis: life in the real world *Mycologist,* Volume 19, Part 3 103-112.

TAYLOR AFS, MARTIN F, READ DJ (2000). Fungal diversity in ectomycorrhizal communities of Norway spruce (*Picea abies* (L.) Karst.) and Beech (*Fagus sylvatica* L.) along north-south transects in Europe. In: Schulze ED, ed. Carbon and nitrogen cycling in European forest ecosystems. Berlin: Springer, 343-365.

TEDERSOO L, SUVI T, LARSSON E, KÕLJALG U (2006) Diversity and community structure of ectomycorrhizal fungi in a wooded meadow. *Mycological Research* IIO(2006) pp. 734-748.

TER BRAAK, C. J. F. (1986). Canonical correspondence analysis: a new eigenvector technique for multivariate direct gradient analysis. Ecology 67:1167-1179.

TIMONEN S, JØRGENSEN KS, HAAHTELA K, SEN R (1998). Bacterial community structure at defined locations of *Pinus sylvestris–Suillus bovinus* and *Pinus sylvestris–Paxillus involutus* mycorrhizospheres in dry pine forest humus and nursery peat. *Canadian Journal of Microbiology* 44: 499–513.

UNESTAM T, STENSTRÖM E (1989).A method for observing and manipulating roots and root-associated fungi on plants growing in nonsterile substrates. *Scandinavian Journal Forest Research* 4:51-58.

UNESTAM T (1991). Water repellency, mat formation, and leaf-stimulated growth of some ectomycorrhizal fungi. *Mycorrhiza*, 1: 13-20.

UNESTAM T, SUN YP (1995). Extramatrical structures of hydrophobic and hydrophilic ectomycorrhizal fungi. *Mycorrhiza* 5: 301-311.

WERNER, A., M. ZADWORNY AND K. IDZIKOWSKA (2002) Interaction between *Laccaria laccata* and *Trichoderma virens* in co-culture and in the rhizosphere of *Pinus sylvestris* grown in vitro. *Mycorrhiza* 12: 139-145.

Month	Cunevo 2005			Cunevo 2006		
	Tmax	Tmin	Pmm	Tmax	Tmin	Pmm
Mai	22.9	9.7	85	29.1	9.4	62.2
June	26.8	13.6	39.5	26.8	12.5	35
October	15.6	6.7	139.5	18.1	7.7	50.6

Tab.1: Temperatures and precipitation in the sites [ISMA (2007); Tmax = Maximum Temperature; Tmin= Minimum Temperature; Pmm= precipitation (mm)].

Fungal taxa	Abbrev	Exploration types	Hydrophobicity
Amphinema sp. (EDM50)		MD fr	hydrophobic
Boletaceae (EDM51)	Bol1	LD	hydrophilic
Boletus sp. (EDM13)	Bolrodo	LD	hydrophobic
Byssocorticium atrovirens (EDM17)	Byssatr	SD	hydrophobic
Cenococcum geophilum (EDM1)	Cenoc	SD	hydrophilic
Cortinarius inochlorus (EDM27)	Corinoc	MD fr	hydrophobic
Cortinarius (EDM57)sp.	Cor1	MD fr	hydrophobic
Cortinarius bolaris (EDM12)	Corbol	MD fr	hydrophobic
Cortinarius cinnabarinus (EDM5)	Corcinn	MD fr	hydrophobic
Cortinarius infractus (EDM62)	Corinfr	MD fr	hydrophobic
Craterellus sp. (EDM41)	Cratell	C/SD	hydrophilic
EDM47	An47	SD	hydrophilic
EDM65	EDM65	MD fr	hydrophobic
EDM68	EDM68	SD	hydrophobic
Entoloma sp. (EDM36)	Entol2	MD sm	hydrophobic
Fagirhiza entolomoides (EDM8)*	Entol1	MD sm	hydrophilic
Fagirhiza arachnoidea (EDM61)	Faracnoid	SD	hydrophobic
Fagirhiza byssoporoides (EDM55)*	Fbyssopo	MD sm	hydrophobic
Fagirhiza cystidiophora (EDM33)	Fcystid	SD	hydrophilic
Fagirhiza fusca (EDM40)	Ffusca	SD	hydrophilic
Fagirhiza lanata (EDM29)	Flanata	MD sm	hydrophilic
Fagirhiza oleifera (EDM2)	Foleifer	C/SD	hydrophilic
Fagirhiza pallida (EDM25)	Fpallida	SD	hydrophilic
Fagirhiza setifera (EDM12)	Fsetif	SD	hydrophilic
Fagirhiza spinulosa (EDM3)	Fspinul	SD	hydrophilic
Fagirhiza stellata (EDM21)*	Tom3	MD sm	hydrophobic
Fagirhiza vermiculiformis (EDM42)	Fvermi	MD sm	hydrophilic
Genea hyspidula (EDM32)	Geneah	SD	hydrophilic
Hydnum sp. (EDM37)	Hydnum	MD fr	hydrophobic
Hygrophorus sp. (EDM26)	Hygro1	C	hydrophilic
Hygrophorus penarius (EDM60)*	Hygro2	SD	hydrophilic
Inocybe sp. (EDM22)	Inoc1	SD	hydrophilic
Laccaria sp. (EDM23)	Lacc	MD sm	hydrophilic

Lactarius acris (EDM56)	Lacacris	MD sm	hydrophilic
Lactarius pallidus (EDM6)	Lpallid	MD sm	hydrophilic
Lactarius rubrocinctus (EDM53)	Lrubroci	MD sm	hydrophilic
Lactarius sp. (EDM48)	Lacta1	C	hydrophilic
Lactarius subdulcis (EDM4)	Lsubdul	MD sm	hydrophilic
Lactarius vellereus (EDM45)	Lvell	MD sm	hydrophilic
Pezizales (EDM 67)	Pezi1	SD	hydrophilic
Piloderma croceum EDM14)	Piloder	MD fr	hydrophobic
Ramaria aurea (EDM43)	Ramaur	MD mat	hydrophobic
Ramaria sp.(EDM58)	Ram2	MD mat	hydrophobic
Ramaria sp. (EDM10)	Ram1	MD mat	hydrophobic
Russula illota (EDM28)	Rusill	C	hydrophobic
Russula mairei (EDM31)	Rusma	C	hydrophilic
Sebacina sp.(EDM34)	Seba2	SD	hydrophilic
Sebacinaceae (EDM11)	Seba1	SD	hydrophilic
Thelephoraceae (EDM63)	Teleph1	MD sm	hydrophobic
Thelephoraceae (EDM66)	Teleph2	MD sm	hydrophobic
Thelephorales (EDM64)	Toml2	MD fr	hydrophobic
Thelephorales (EDM59)	Tomlo1	MD fr	hydrophobic
Tomentella sp. (EDM18)	Tom1	MD fr	hydrophobic
Tomentella sp. (EDM19)	Tom2	MD sm	hydrophilic
Tomentella sp.(EDM46)	Tom4	MD sm	hydrophilic
Tomentella sp.(EDM70)	Tom5	SD	hydrophilic
Tricholoma acerbum (EDM24)	Tricacer	MD fr	hydrophobic
Tricholoma sciodes (EDM39)	Tricscio	MD fr	hydrophobic

Tab. 2.: **Exploration types of the anatomotypes and the relationship with the hydrophobicity [C= contact type - SD= short distance; MD sm= medium distance smooth; MD fr= medium distance fringe; MD mat= medium distance mat; LD= long distance].**

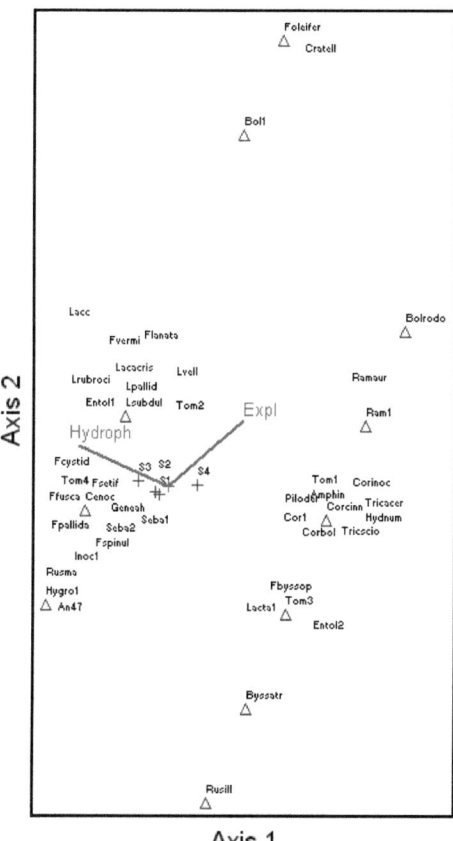

Fig. 1: CCA joint biplot of the EM fungal community in the S1, S2, S3, S4 stands in June 2005 (crosses). Open triangles represent the EM species. Vectors indicate the ecological features as quantitative parameters: the hydrophibicity attitude according to Unestam 1991 and the exploration type i.e. the potential exploration strategies in the soil according to Agerer 2001 [Hydroph = hydrophobicity; Expl = Exploration types; see Tab. 2].

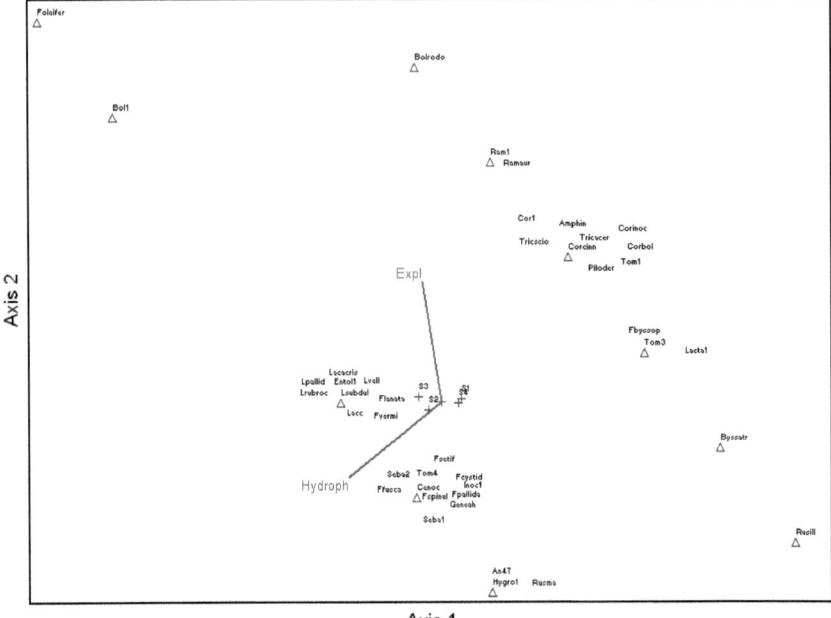

Fig. 2: CCA joint biplot of the EM fungal community in the S1, S2, S3, S4 stands in October 2005. Open triangles represent the EM species. Vectors indicate the ecological features as quantitative parameters [Hydroph = hydrophibicity; Expl = Exploration types].

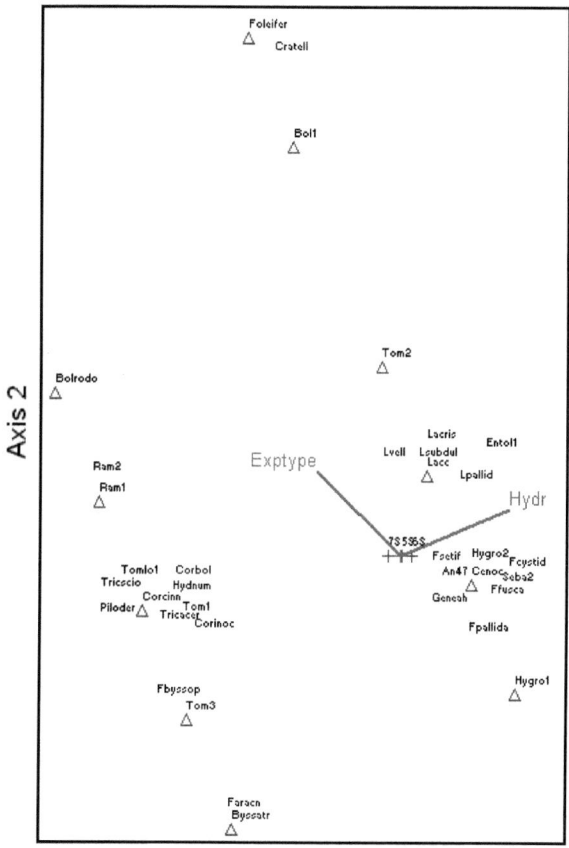

Fig. 3: CCA joint biplot of the EM fungal community in the S5, S6, S7 stands (crosses). Open triangles represent the EM species. Vectors indicate the ecological features as quantitative parameters [Hydroph = hydrophibicity; Expl = Exploration types].

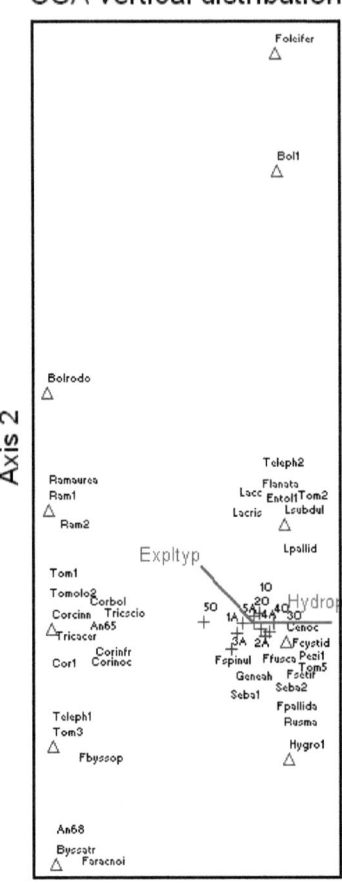

Fig. 4: CCA joint biplot of the EM fungal community in the S1, S2, S3, S4, S5 stands. Open triangles represent the EM species. Vectors indicate the ecological features as quantitative parameters [Hydroph = hydrophibicity; Expl = Exploration type; O, A : soil horizons; f. e. 5O= site S5 O organic horizon].

CHAPTER 5

"Fagirhiza entolomoides" + Fagus sylvatica (L.)

ERIKA DI MARINO[1,2], LUCIO MONTECCHIO[1], REINHARD AGERER[3],

[1]Università degli Studi di Padova Agro-Forestry Systems and Land Use (TeSAF) Department V.le dell'Università, 16 - 35020 Legnaro (PD) – Italy

[2]Centro di Ecologia Alpina - Viote del Monte Bondone (TN), Località Caserme, 2 38060 Garniga Terme (TN)- Italy

[3]Organismic Biology: Mycology, Department Biology and GeoBio-Center^{LMU}, University of München, Menzinger Str., 67, D-80638, Germany

Short description

The ectomycorrhizae are characterized by very long mycorrhizal systems (up to 7 cm), with sinuous or tortuous, woolly, brownish pink mycorrhizal tips when younger, whitish or colourless when older. The root is frequently shining through the mantle. The outer mantle is loosely plectenchymatous with very wide hyphae of (5)6-(8) μm diam. that are arranged in parallel bundles and possess clamps. The middle mantle is similar to the other, with broad streaks of parallel hyphae of 5-7 μm diam. The inner mantle has also broad streaks of parallel hyphae, but they are with 4-5 μm diam. slightly thinner. Rhizomorphs are undifferentiated, very frequent, compact and are formed as stout, short, conical structures, (2)5-12(65) μm wide, rarely up to 90 μm. Sometimes a loose gelatinous matrix is visible on the surface of the outer mantle and on the rhizomorphs. The Hartig net is not uniform and is similar to that of other typical ectomycorrhizal species of the genus *Entoloma*. In particular as it is patchily distributed, but it is paraepidermal where present.

Morphological characters (Figs. 1): *Mycorrhizal systems* irregular monopodial-pyramidal, up to 70 mm long; with lots of stout rhizomorphs appearing as short, slenderly conical structures,hydrophilic, smooth subtype of medium-distance exploration type. - *Main axes* up to 0.5-1 mm diam., tortuous and sinuous. - *Unramified ends* (0.5)4(5) mm long and 0.25 mm diam., not inflated, cylindric or tapering, whitish-pink due to root colour, brownish-pink when younger; distinct mantle surface visible, with semi-transparent mantle when older; cortical cells visible through older tips, not carbonizing, dots, cystidia and emanating hyphae lacking. - *Surface of unramified ends* loosely stringy, densely or loosely woolly. - *Rhizomorphs* not differentiated, frequent, round or nearly so in cross-section, concolorous to the mantle, colourless, pinkish white, connection to the mantle distinct (Fig.1), distribution not specific; margin of rhizomorphs smooth. Lacking are nodia. - *Sclerotia* not found.

Anatomical characters of mantle in plan views: Lacking are cells densely filled with oily droplets or cells homogeneously filled with brownish contents, blue granules, needle-like contents, drops of exuded pigment. - *Outer mantle layers* (Fig. 2a): plectenchymatous, hyphae arranged in parallel bundles, but no special pattern discernible (mantle type B, according to according to AGERER 1991, 1995, AGERER 1987-2006, AGERER & RAMBOLD 2004-2007); hyphae cylindric and constricted at septa or slightly inflated at middle portions, (5)6(8) µm diam., cells (11)35-60(70) µm long, smooth, with clamps, hyphae colourless or membranaceously slightly yellowish, walls 0.1-0,5 µm thick, septa as thick than walls; a slightly gelatinous matrix present . - *Middle mantle layers* (Fig. 3a): plectenchymatous, with broad streaks of parallel hyphae; cells colourless, smooth, 5-7 µm diam., matrix lacking, cell walls up to 0.1-0.5 µm; anastomoses infrequent, open. - *Inner mantle layers* (Fig. 3b): plectenchymatous with broad streaks of parallel hyphae, hyphae 4-5 µm diam., hyphal portions (1)3-4(5) µm long, cell walls up to 0.1-0.5 µm, matrix lacking.

Anatomical characters of emanating elements (Figs. 4): Lacking are gelatinized hyphae, drops of exuded pigment, and in IC strongly light reflecting crystals, internal nodia, and conical structures. - *Rhizomorphs* (Figs. 4) (2)5-12(65) µm diam., exceptionally up to 90 µm, undifferentiated, type A/B (according to AGERER 1991, 1995, AGERER 1987-2006, AGERER & RAMBOLD 2004-2007, AGERER & IOSIFIDOU 2004), hyphae of uniform diam., or slightly inflated at septum; central hyphae 2-4 µm diam., cell walls 0.2-0.5 µm, pores indistinct, septa with the same thickness as walls; cells 5-60 µm long, colourless or membranaceously yellowish; sometimes surface covered by a slightly gelatinous matrix, infrequently ramified (Fig. 2b). - *Emanating hyphae* not observed. – *Cystidia* not observed. - *Chlamydospores* not observed.

Anatomical characters, longitudinal section: Mantle (30)40-70(100) μm wide, at very tip 25-70 μm, plectenchymatous, different layers not discernible, but at points of the connection to rhizomorphs regular organization lacking; hyphae of the unlayered mantle tangentially (3)20-30(40) μm, radially (3)4-7(8) μm. - *Tannin cells* lacking, with calyptra cells. – *Epidermal cells* with Hartig net paraepidermal, not homogeneously distributed over the section, in part nest-like, i.e. only a few neighbouring epidermal cells with Hartig net present, cells rectangular or tangentially-oval to – elliptic, oriented parallel to the root, tangentially (40)60-80(140) μm, radially (15)20-30(35) μm. - $EC_t = 68$, $EC_q = 2,9$. - *Hartig net in section* hyphal cells roundish to cylindrical, in one row, (2)3-4(6) μm thick; haustoria lacking. *Hartig net in plan view* a slightly ramified palmetti-type without septa, infrequently lobed, often only hypha-like and clamps visible, lobes (1.5)2-3(4) μm, broad.

Colour reaction with different reagents: Mantle and rhizomorph preparations: KOH 10%: n. r. (no reaction); cotton blue: slightly bluish; ethanol 70%: n. r.; FEA: n .r.; iron (II) sulphate: slightly bluish; lactic acid: n. r.; Melzer's reagent: n.r.; sulpho-vanillin: mantle and rhizomorphs slightly reddish.

Autofluorescence: Whole mycorrhiza: UV 254 nm: lacking; UV 366 nm: lacking. - *Mantle in section:* UV-filter 340-380 nm: very slightly bluish; blue-filter 450-490nm: very slightly yellowish; green filter 530-560 nm: very slightly reddish. – *Rhizomorph in section:* UV-filter 340-380 nm: slightly bluish, margin stronger; blue filter 450-490 nm: slightly yellowish, margin stronger; green-filter 530-560 nm: n. r.

Reference specimen: Italy, province Trient (Trentino-Alto Adige Region), Val di Non, district Denno (46°14' N; 10°57' E), beech coppice, more frequent in organic layers, 5.06.2005, myc. isol E. Di Marino, EDM 8 in FEA (in PD). – *Additional specimens examined*: Italy, province Trient (Trentino-Alto Adige Region), Val di Non, district Denno (46°14' N; 10°57' E), beech coppices, 1050-1200 m a.s.l., June 2005, EDM 8a in FEA (in PD), October 2005 EDM 8b in FEA (in PD), May/June 2006, EDM 8c in FEA (in PD). – Soil conditions for all collections: mesic or xeric, pH of the soil 6-6.6, N_{tot} 6,7-15,9, C/N 17-18, C_{org} 111-279g/Kg; in mineral layers pH 5.2-6.5, N_{tot} 11,5-26,3, C_{org} 259-434 g/Kg, C/N 16-19.

DNA analyses: DNA-analyses, sequence evaluation and alignment were performed (EDM 8c) according to Tedersoo et al. (2006), best match in Unite 91% with UDB000937 *Entoloma* sp., 97% in NCBI BLASTn search in GenBank Uncultured ectomycorrhiza (Entolomataceae) AJ938003 18S rRNA gene (partial), 5.8S rRNA gene, 28S rRNA gene (partial), ITS1 and ITS2. Similarity of 86% with *Entoloma sinuatum* isolate AFTOL-ID 524 DQ486700 internal transcribed spacer 1, 5.8S ribosomal RNA gene, and internal transcribed spacer 2, complete sequence, and query coverage of 95%. GenBank Accession number EU444549.

Discussion:

This unidentified ectomycorrhiza is very similar to that of *E. nitidum* on *Carpinus betulus* (MONTECCHIO et al. 2006), as both are characterized by stout, conical rhizomorph-like structures on the surface of the mantle with no specific origin, and by epidermal cells that are visible through the mantle. The very long and loosely branched systems are also characteristic for both species and are exceptional for ectomycorrhizal systems (Agerer & Rambold 2004-2007). We therefore conclude that *F. entolomoides* is very likely formed by a species of the genus *Entoloma*. Differences between the two species regard the absence of a matrix and the occurrence of thicker cell walls in mantles and of thinner rhizomorphs in *Entoloma nitidum*. In addition, the hyphal orientation in the inner mantle layers of *E. nitidum* ectomycorrhizae is irregular or parallel, whereas in *F. entolomoides* it is always in parallel. *Entoloma sinuatum* on *Salix* (AGERER 1997; 1998) forms plectenchymatous middle mantle layers intermixed with pseudoparenchymatous portions, what is unknown in *E. nitidum* and *F. entolomoides*. Stout, conical rhizomorph-like structures are lacking in *E. sinuatum*, whereas undifferentiated rhizomorphs occur instead, accompanied by many emanating hyphae. The Hartig net of *E. sinuatum* is patchy, too, as in *F. entolomoides*. A matrix is only rarely visible on the rhizomorphs or on the mantle surface of *F. entolomoides*, as compared to *E. sinuatum* (AGERER 1997).

Other species, *Entoloma alpicola* (J. Favre) Bon & Jamoni (GRAF & BRUNNER 1996), *E. erophilum* (Fr.) P. Karst. (ZEROVA & ROZHENKO 1966), *E. rhodopolium*(Fr.) P. Karst. (MODESS 1941), and *E. sericeum* (Bull.) Quél. (ANTIBUS et al. 1981) are too briefly characterized for a comparison to the above mentioned ectomycorrhizae.

A very peculiar situation is the parasitic behaviour of *Entoloma clypeatum* f. *hybridum* on *Rosa multiflora* (KOBAYASHI & HATANO 2001) and of *Entoloma saepium* (Noulet & Dass) Richon & Roze on *Rosa* sp. and *Prunus* sp. (AGERER & WALLER 1993, AGERER 2006) that digests the root meristem. At least morphologically this mycorrhiza is very similar to that of *Entoloma clypeatum* on *Prunus cerasus* (ANDRUSZEWSKA & DOMINIK 1971).

In conclusion, the main features of *F. entolomoides* are similar to those reported for other species that belong to the genus *Entoloma*. But DNA-sequencing confirms only a similarity of of 91% and

97% by BLASTn search in UNITE and in GenBank, respectively. This is the first description of an ectomycorrhiza of the genus *Entoloma* on *Fagus* (DE ROMAN et al. 2005; AGERER & RAMBOLD 2004-2007).

Acknowledgments: We thank to Dr. S. Raidl, Dr. P. Bodensteiner, Nourou Soulemane Yorou, Dr. L. Beenken, R. Verma, Miss C. Bubenzer-Hange, Mr. E. Marksteiner for discussions and technical help, the "A. Gini Foundation" of Padua University for the partial financial support, and the Centre of Alpine Ecology of Monte Bondone (TN-Italy) for soil analysis.

References: AGERER R (1987-2006) Colour atlas of ectomycorrhizae. 1th -13 th del. Einhorn, Schwäbisch Gmünd, Germany. - AGERER R (1991) Characterization of ectomycorrhiza. In Norris JR, Read DA, Varma AK (eds) Techniques for the study of mycorrhiza. Methods in Microbiology, vol 23, Academic Press, London et al., pp 25-73. - AGERER R (1995) Anatomical characteristics of identified ectomycorrhizas: an attempt towards a natural classification. In Varma K, Hock B (eds) mycorrhiza: structure, function, molecular biology and biotechnology. Springer, Berlin, Heidelberg, New York, pp 685-734. - AGERER R (1997) *Entoloma sinuatum* (Bull.: Fr.) Kummer + *Salix* spec. Descr Ectomyc 2: 13-18. - AGERER R (1998) *Entoloma sinuatum*. In Agerer R (ed) Colour Atlas of Ectomycorrhizae, plate 117, Einhorn-Verlag, Schwäbisch Gmünd. - AGERER R (2006) Fungal relationship and structural patterns of their ectomycorrhizae. Mycological Progress 5(2): 67-107. - AGERER R, RAMBOLD G (2004-2007) [first posted on 2004-06-01; most recent update: 2007-03-02]. DEEMY – An Information System for Characterization and Determination of Ectomycorrhizae. www.deemy.de – München, Germany. - AGERER R, IOSIFIDOU P (2004) Rhizomorph structures of Hymenomycetes: a possibility to test DNA-based phylogenetic hypotheses? In Agerer R, Piepenbring M, Blanz P (eds) Frontiers in Basidiomycote Mycology, IHW-Verlag, Eching, pp 249-302. - AGERER R, WALLER K (1993) Mycorrhiza of *Entoloma saepium:* parasitism or symbiosis? Mycorrhiza (1993) 3: 145-154. - ANDRUSZEKWSKA A, DOMINIK T (1971) Mycorrhiza of *Prunus cerasus* L.x *Rhodophyllum clypeatus* (Bull. Ex Fr.) Quél. Zeszyty Nauk Wyzsz Szkoly Roln Szczecin 37:3-14. - ANTIBUS RK, CROXDALE JG, MILLER OK, LINKINS AE (1981) Ectomycorrhizal fungi of *Salix rotundifolia* III. Resynthesized mycorrhizal complexes and their surface phosphatase activities. Can J Bot 59: 248-2465. - DE ROMAN M, CLAVERIA V, DE MIGUEL AM (2005) A revision of the description of ectomycorrhiza published since 1961. Mycol. Res. 109 (10):1063-1104. - GRAF F, BRUNNER I (1996) Natural and synthesized ectomycorrhizas on the alpine dwarf willow

Salix erbacea. Mycorrhiza 6: 227-235. – KOBAYASHI H, HATANO K (2001) A morphological study of the mycorrhiza of *Entoloma clypeatum* f. *hybridum* on *Rosa multiflora*. Mycoscience 42: 83-90. - - MODESS O (1941) Zur Kenntnis der Mykorrhizabildner von Kiefer und Fichte. Symb Bot Upsal 5(1): 1-146. - MONTECCHIO L, ROSSI S, GARBAYE J (2006) *Entoloma nitidum* Quèl + *Carpinus betulus*. Descr Ectomyc 9/10: 33-38. -TEDERSOO L, SUVI T, LARSSON E, KÕLJALG U (2006) Diversity and community structure of ectomycorrhizal fungi in a wooded meadow. Mycol. Res. 110(2006) 734-748. - ZEROVA MY, ROZHENKO GL (1966) *Entoloma erophilum* (Fr.) P. Karst. t. and *E. serieum* (Bull.) Quèl. Mycorrhizal symbionts of the oak. Ukr Bot Zh 23:87-90.

Captions: **Figs. 1.** - **a, b, c.** Habit. - **Fig. 2** - **a.** Plan view of outer mantle layer. – **b.** Older conical rhizomorph-like structures with ramification. - **Fig. 3** - **a.** Middle mantle layer with broad streaks of parallel hyphae, open anastomosis (asterisk) and clamps. –**b.** Inner mantle layer with broad streaks of parallel hyphae. - **Fig. 4.** Differently thick conical rhizomorph-like structures; 'a' in surface view (above) and optical section (below). - **Fig. 5.** Connection point of conical rhizomorph-like structures to the outer mantle where the hyphal structure is ring-like. *All figs. from EDM 8a (in PD).*

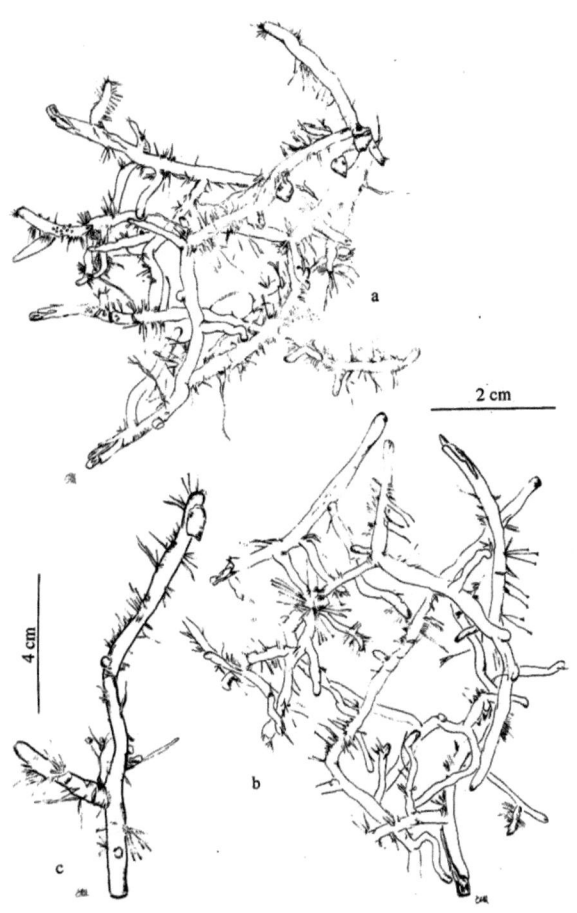

Figs. 1a, 1b, 1c.

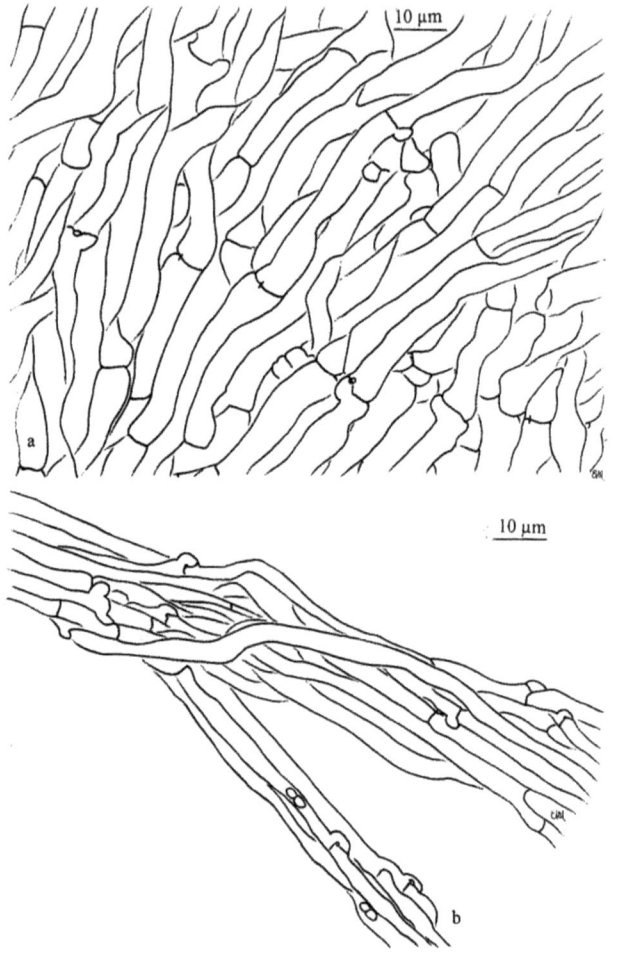

Figs. 2a, 2b.

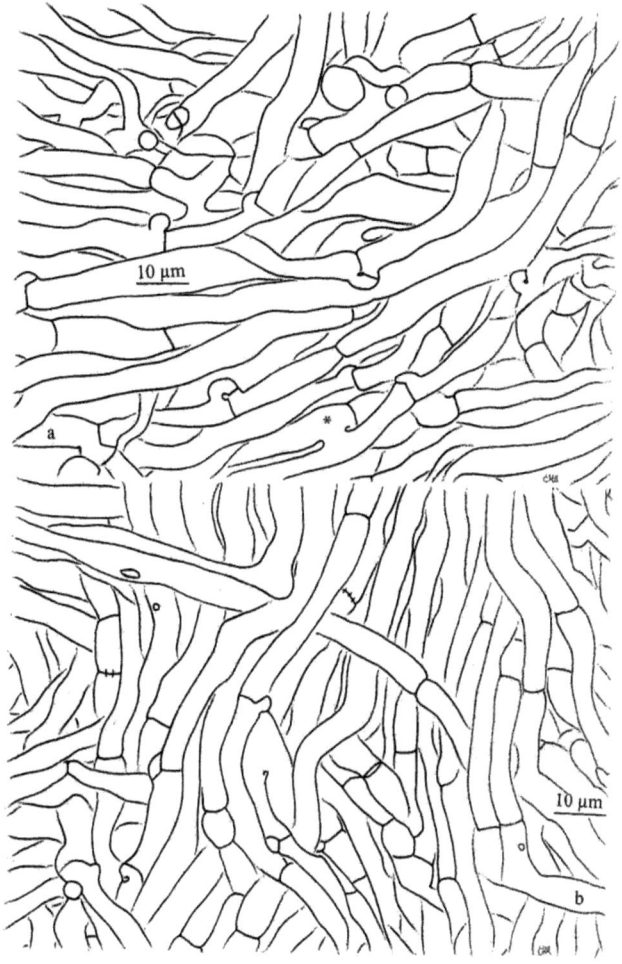

Figs. 3a, 3b.

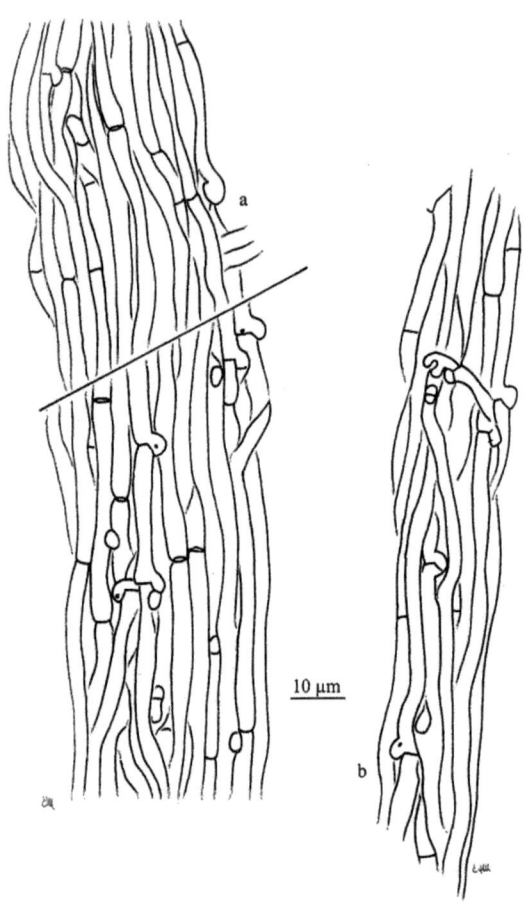

Figs. 4a, 4b.

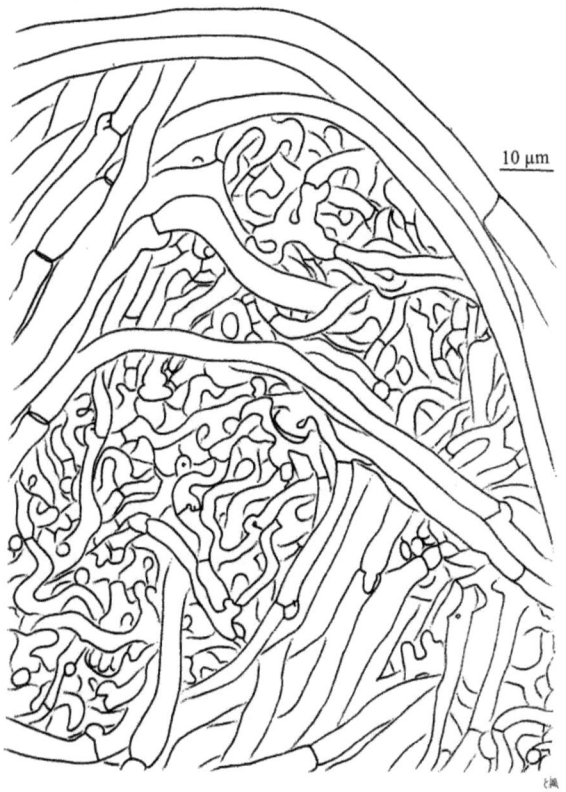

Fig. 5.

CHAPTER 6

"Fagirhiza byssoporoides" + Fagus sylvatica (L.)

ERIKA DI MARINO[1,2], LUCIO MONTECCHIO[1], REINHARD AGERER[3]

[1] Università degli Studi di Padova Agro-Forestry Systems and Land Use (TeSAF) Department V.le dell'Università, 16 – I-35020 Legnaro (PD) – Italy

[2] Centro di Ecologia Alpina - Viote del Monte Bondone; I-38070 Sopramonte (TN), Italy.

[3] Organismic Biology: Mycology, Department Biology and GeoBio-Center^{LMU}, University of München, Menzinger Str., 67, D-80638, Germany

Short description

The ectomycorrhizae are irregularly pinnate or irregularly dichotomous, whitish greyish, when young loosely cottony, with a deep blue bruising reaction on the surface of rhizomorphs and mantle. When older, the blue zone becomes more distinct, and the ectomycorrhiza is woollier. The diagnostic characteristics are a plectenchymatous mantle with a very thick gelatinous matrix, octahedral crystals on the surface, and infrequent staghorn-shaped hyphae on the mantle surface. Rhizomorphs show a typical differentiation with closely packed, straight, parallel hyphae in the centre with infrequent clamps and a thin matrix, sheathed by a thin layer of narrow, entwining, thin-walled, rarely simple septate, clamp-less hyphae with a slightly gelatinous layer that occur together with some very infrequently branched hyphae; outer portions of the rhizomorphs with irregular crystals. The main features of this ectomycorrhiza are similar to those reported for ectomycorrhizae of the genus *Byssoporia*.

Morphological characters (Figs. 1): *Mycorrhizal systems* with 2-3 orders of ramification, (2)5-14(18) side-branches per cm; hydrophobic, smooth subtype of medium-distance exploration type. - *Main axes* up to 8 mm long and 0.5 mm diam. - *Unramified ends* up to 1.7(2) mm long and 0.25 mm diam., not inflated, straight and rarely bent, cylindric, with rounded tips, and with a constricted bluish base, whitish greyish with bluish spots; older parts more distinctly blue in patches. – *Surface of unramified ends* loosely cottony or woolly, very tip loosely cottony, with few emanating hyphae. – *Rhizomorphs* abundant in older mycorrhizal systems, whitish, quite compact, frequently and repeatedly branched, with hairy to woolly surface, originating at the very base of mycorrhizal systems and at the very tips, connecting different systems; round in cross-section, turning bluish-violet; concolourous to the mantle; distinct connection with the mantle. – *Sclerotia* not observed.

Anatomical characters of mantle in plan views (Figs. 2, 5b): *Cells without contents and hyphae without clamps.* - *Mantle surface* with a thin to very thick gelatinous matrix between hyphae, with crystals; crystals octahedral to bipyramidal or acicular, regularly shaped crystals (2)4-7(8) µm, acicular crystals (2)4-11(13) µm long, with irregularly shaped, multiply branched cystidia-like hyphae of 1.5-3 µm diam., 15-20 µm long, without septa. - *Outer mantle layers* (Figs. 2) plectenchymatous, with slightly ring-like pattern and gelatinous, sometimes very thick matrix (mantle type A/C, according to AGERER 1991, 1995, AGERER 1987-2006, AGERER & RAMBOLD 2004-2007); hyphae cylindric, not constricted at septa, irregularly shaped; simple septate; angles between hyphal junctions ca. 90-120°, membranaceously and plasmatically bluish due to bruising reaction, otherwise colourless, smooth, cell walls 0.2-0.5 µm; cells (2)4-5(6) µm diam., 20-25 µm long. - *Middle mantle layers* (Fig. 3a) plectenchymatous, slightly ring- to star-like, with infrequent simple septa, membranaceously and plasmatically bluish-violet due to bruising reaction, otherwise colourless, cell walls 0.2-0.5 µm, smooth, cells (2)3-5(8) µm diam., 15-20 µm long; with (2)4-5(8) µm large crystals. - *Inner mantle layers* (Fig. 3b) plectenchymatous, with occasionally ring-like arranged hyphae and gelatinous matrix, bruising reaction visible, too; all hyphae irregularly shaped, with infrequent simple septa, cells 3-4 µm., distance of hyphal septa 2.5-3.5 µm.

Anatomical characters of emanating elements (Figs. 4, 5a, 6, 7): Lacking are gelatinized hyphae, ampullate hyphae, drops of secreted pigment, and in IC strongly light reflecting crystals, nodia and conical structures on rhizomorphs; cell walls smooth. - *Rhizomorphs* (Figs. 4, 5a, 6, 7) of type C (AGERER 1991, 1995, AGERER 1987-2006, AGERER & RAMBOLD 2004-2007; AGERER 1999; AGERER & IOSIFIDOU 2004), (20) 40-70(100)µm diam.; when young undifferentiated (Fig. 4), with open anastomoses, without clamps, cells of the peripheral hyphae occasionally irregularly shaped; thicker rhizomorphs covered with thin peripheral hyphae (Fig. 5a, 6,7), density increasing with thickness of rhizomorphs, below them with small irregularly shaped crystals; peripheral hyphae 2-3 μm diam., some of them ramified and of cystidia-like shape, without clamps, rarely with simple septa, with slightly gelatinous surface, central hyphae somewhat enlarged with a slight gelatinous matrix, weakly inflated at the septa, with infrequent clamps, simple septa abundant, distance of septa (8)15-40 μm, cell walls 0.2-0.5 μm. - *Emanating hyphae* not frequent, tortuous, few irregularly branched, with slightly gelatinous surface, similar to those on mantle and rhizomorph surface, with simple septa, 2-2.5 μm diam. – *Chlamydospores* not observed.

Anatomical characters, longitudinal section: Mantle compressed and thin, 10-30 μm wide, very tips with 10(15)-20 μm thick mantle, twisted tips very frequent; different layers discernible; outer, middle and inner mantle layers plectenchymatous; outer mantle layer hyphae tangentially 2-4 μm and radially (8)10-12(15) μm; middle mantle layer hyphae tangentially 3-4 μm and radially (8)10-12(15) μm, inner mantle layer hyphae tangentially 3-5 μm and radially 3-4 μm. – *Tannin cells* lacking. - *Epidermal cells* rectangular, tangentially (30)35-40(50) μm and radially (10)12-20(25) μm; EC_t = 38, EC_q = 0,4. - *Hartig net in section* paraepidermal; *Hartig net in plan view* of palmetti-type, lobes without septa, lobes 1-2 μm broad.

Colour reaction with different reagents: Mantle and rhizomorph preparations: cotton blue: slightly bluish; ethanol 70%: n.r.; FEA: n.r.; iron(II)sulphate: crystals dissolving; KOH 15%: deep blue pigment disappearing; lactic acid: blue pigment dissolving , crystals slowly dissolving; Melzer's reagent: n.r.; sulpho-vanillin: n.r., but mantle and rhizomorphs slightly rosy or reddish.

Autofluorescence: Whole mycorrhiza: UV 254 nm: lacking; UV 366 nm: lacking; -*Mantle in section:* UV-filter 340-380 nm: bluish; blue-filter 450-490nm: yellowish; green filter 530-560 nm: reddish. – *Rhizomorph in section:* UV-filter 340-380 nm: slightly bluish, margin stronger; blue filter 450-490 nm: slightly yellowish, margin stronger; green-filter 530-560 nm: slightly red.

Reference specimen for ectomycorrhiza: Italy, province Trient (Trentino-Alto Adige Region), Val di Non, district Denno (46°14' N; 10°57' E), beech coppice, 1050-1200 m a.s.l.; myc. isol E. Di Marino, 20.10.2005, EDM 55 in FAA in PD. – *Additional specimens examined*: Italy, province Trient (Trentino-Alto Adige Region), Val di Non, district Denno (46°14' N; 10°57' E), beech coppices of different ages, 1050-1200 m a.s.l.; myc. isol E. Di Marino, 20.10.2005, EDM 55a in FAA in PD. – Soil conditions for all collections: more frequent in mineral layers, mesic or xeric, pH of the soil 5,1-6,7, N_{tot} 3,8-15,6, C/N 15-18, C_{org} 61-241 g/Kg; in organic layers pH 4.2-5.6, N_{tot} 20-

22.6, C_{org} 361-392 g/Kg, C/N 17-18.

DNA analyses: From the mycorrhizal root tips of *F. byssoporioides* obtained DNA was amplified and sequenced using the primers ITS1-F and ITS4. The applied methods of DNA extraction, PCR, and sequencing follow TEDERSOO et al. (2006). The PCR product of the targeted nuclear ITS rDNA (complete sequence of internal transcribed spacer 1, 5.8S ribosomal RNA gene, and internal transcribed spacer 2, flanked by partial sequences of 18S and 28S ribosomal RNA genes) has a size of 594bp. The GenBank accession number of *F. byssoporioides* ECM is EU444550.

A BlastN search was performed in GenBank using the sequence of the *F. byssoporioides* ECM as query. The retrieved matches had a maximum sequence identity of 87% at best. The best matches belong to samples of unidentified ectomycorrhizal fungi representing members of Leucogastraceae, Albatrellaceae, and Agaricaceae as well as several isolates of *Leucophleps spinispora*. In the context of anatomy based similarities to *Byssoporia* the *F. byssoporioides* ECM sequence was compared to an unpublished ITS sequence (complete sequence of internal transcribed spacer 1 and 5.8S ribosomal RNA gene, partial sequence of internal transcribed spacer 2; flanked by partial sequence of the 28S ribosomal RNA gene) with a length of 553bp. The latter sequence was generated from a *Byssoporia terrestris* fruitbody on *Picea abies* from Germany (SR 1101, in herb. S. Raidl) the associated ECM of which has been described by SCATTOLIN et al. (2006). Both sequences are nearly identical showing one differing base in the internal spacer region 1. Additionally, the BlastN search in GenBank using the *F. byssoporioides* ECM sequence as query retrieved the sequence from a *B. terrestris* fruitbody from Sweden (Hjm 18172, in herb. GB; accession number EU118608, see LARSSON 2007) with an identity value of 84% (query coverage 100%). The only *Byssoporia* sequence (UDB001766) deposited in UNITE (see KÕLJALG et al. 2005) was generated from the same voucher specimen of *B. terrestris*. It includes the partial sequence of the 5.8S ribosomal RNA gene and the complete sequence of internal transcribed spacer 2 that compared with the covered part of the *Fagirhiza byssoporioides* ECM sequence produced an identity value of 91%.

Discussion: This ectomycorrhiza on *Fagus sylvatica* is likely formed by a member of the genus *Byssoporia*, because of the peculiar anatomical features of the rhizomorphs, with a cover of peripheral, twisted hyphae. Up to now ectomycorrhizae of *Byssoporia* species were described only from gymnosperms (DE ROMAN et al. 2005). Five different varieties of *Byssoporia terrestris* + *Pseudotsuga menziesii* have been illustrated and described by ZAK (1969) and ZAK & LARSEN (1978). These five varieties are suggested to belong to three different species due to the diversity of peripheral hyphae (AGERER 2006). According to the description of the ectomycorrhizae of the different varieties of *Byssoporia terrestris* (ZAK 1969, ZAK & LARSEN 1978), *B. terrestris* var. *sublutea* M.J. Larsen & Zak is the closest to *Fagirhiza byssoporioides*, as in both ectomycorrhizae the rhizomorphs are covered by strongly twisted, even corkscrew-like peripheral hyphae. Contrary to that variety, but, similarly to *B. terrestris* var. *sartoryi* (Bourdot & L. Maire) M.J. Larsen & Zak and *B. terrestris* var. *lilacinorosea* M.J. Larsen & Zak, *F. byssoporioides* reveals some staghorn-like hyphae on the mantle surface and between the twisted peripheral rhizomorph hyphae. They do not form, however, a homogeneous cover as in the latter varieties. The ectomycorrhizae of *B. terrestris* described by SCATTOLIN et al. (2006) that could not be identified to variety-level, form typical cork-screw-like peripheral rhizomorph hyphae, more distinctly twisted than in *F. byssoporioides*, have clamps and lack ramified peripheral hyphae. Both, however, turn blue after bruising. In addition, the outer mantle layers differ completely. That of *B. terrestris* sensu SCATTOLIN et al. (2006) is formed by mostly normal hyphae with some irregularly shaped hyphal structures, whereas the mantle of *F. byssoporioides is* exceptionally irregularly shaped with multiply branched cystidia-like hyphae. Regarding these anatomical differences the ECM of *B. terrestris* on *Picea abies* described by SCATTOLIN et al. (2006) and *F. byssoporioides* very likely represent different taxa although the ITS sequences generated from a *B. terrestris* fruitbody (DI MARINO unpublished) that was associated with its ectomycorrhizae on *Picea abies* (SCATTOLIN et al. 2006) and *F. byssoporioides* are nearly identical. Another reason for the – preliminary – classification of the ECM as a *Fagirhiza* is the extreme difference between the sequences from fruitbodies identified as *B. terrestris* in SCATTOLIN et al. (2006) (>99% sequence identity with *F. byssoporioides*) and LARSSON (2007) (84% sequence identity with *F. byssoporioides*), respectively. This discrepancy and the still very limited knowledge about relationships and taxonomy of *Byssoporia* accentuate the need for further detailed studies of this group.

Apart from our collection, a bluish bruising reaction is only known from *B. terrestris* (SCATTOLIN et al. 2006), and *B. terrestris* var. *satoryi* (ZAK 1969). The rhizomorphs of *B. terrestris* var. *satoryi* are completely covered by typical staghorn-shaped hyphae, whereas in *F. byssoporioides* only a few peripheral hyphae are scarcely ramified and intermixed between the twisted hyphae. It can therefore be concluded, that *F. byssoporioides* is distinct from all hitherto characterized ecto-mycorrhizae of the genus *Byssoporia*. Crystals on rhizomorphs and mantles are not reported for either ectomycorrhiza, but ZAK (1969) and ZAK & LARSEN (1978) reported on encrusted central and peripheral hyphae.

The phenomenon of bluing, not necessarily a bruising reaction, as found in the genus *Byssoporia* is well known from some ectomycorrhizae of Boletales (AGERER 2006; AGERER & RAMBOLD 2004-2007): *Alpova diplophleus* (Zeller & Dodge) Trappe & Smith (MILLER et al. 1988, WIEDMER et al. 2001), *Chamonixia caespitosa* Rolland (RAIDL 1999), and three *Leccinum* ectomycorrhizae described by MÜLLER & AGERER (1990): *L. hopolus* (Rostk.) Watling, *L. scabrum* (Bull.: Fr.) S.F. Gray and *L. variicolor* Watling). However, *Byssoporia* ectomycorrhizae do not form highly differentiated rhizomorphs with central vessel hyphae (type F, according to AGERER 1987-2006) that are the main common feature of all ectomycorrhizae of the *Boletales* ss. Agerer (AGERER 1999, 2006, AGERER & IOSIFIDOU 2004).

Acknowledgments: Thanks are due to Dr. Stefan Raidl for the exsiccate, Dr. Ludwig Beenken, Dr. Philomena Bodensteiner, Miss C. Bubenzer-Hange, Mr. E. Marksteiner, and the "A. Gini Foundation" of Padua University for the support, and the Centre of Alpin Ecology of Monte Bondone (TN-Italy) for the soil analyses.

References: AGERER R (1987-2006) Colour atlas of ectomycorrhizae. 1st -13th del. Einhorn, Schwäbisch Gmünd, Germany. - AGERER R (1991) Characterization of ectomycorrhiza. In Norris JR, Read DA, Varma AK (eds) Techniques for the study of mycorrhiza. Methods in Microbiology, vol 23, Academic Press, London et al., pp 25-73. - AGERER R (1995) Anatomical characteristics of identified ectomycorrhizas: an attempt towards a natural classification. In Varma K, Hock B (eds) mycorrhiza: structure, function, molecular biology and biotechnology. Springer, Berlin, Heidelberg, New York, pp 685-734. - AGERER R (1999) Never change a functionally successful principle: the evolution of *Boletales* s.l. (Hymenomycetes, Basidiomycota) as seen from below-ground features. Sendtnera 6: 5-91. - AGERER R (2006) Fungal relationship and structural patterns of their ectomycorrhizae. Mycological Progress 5(2): 67-107. - AGERER R, IOSIFIDOU P (2004) Rhizomorph structures of Hymenomycetes: a possibility to test DNA-based phylogenetic hypotheses? In Agerer R, Piepenbring M, Blanz P (eds) Frontiers in Basidiomycote Mycology, IHW-Verlag, Eching, pp 249-302. - AGERER R, RAMBOLD G (2004-2007) [first posted on 2004-06-01; most recent update: 2007-05-02]. DEEMY – An Information System for Characterization and Determination of Ectomycorrhizae. www.deemy.de, München, Germany. - DE ROMAN M, CLAVERIA V, DE MIGUEL AM (2005) A revision of the description of ectomycorrhiza published since 1961. Mycol. Res. 109 (10)1:1063-1104. - KÕLJALG U, LARSSON K-H, ABARENKOV K, NILSSON RH, ALEXANDER IJ, EBERHARDT U, ERLAND S, HOILAND K, KJOLLER R, LARSSON E, PENNANEN T, SEN R, TAYLOR AFS, TEDERSOO L, VRALSTAD T, URSING BM. (2005) UNITE: a database providing web-based methods for the molecular identification of ectomycorrhizal fungi. New Phytol 166: 1063-1068 - LARSSON K-H (2007) Re-thinking the classification of corticioid fungi. Mycol. Res. 111 (9): 1040-1063. - MILLER SL, KOO CD, MOLINA R (1988) An oxidative blue-bruising reaction in *Alpova diplophloeus* (Basidiomycetes, Rhizopogonaceae) + *Alnus rubra* ectomycorrhizae. Mycologia 80(4): 576-581. - MÜLLER W, AGERER R (1990) Studien an Ektomykorrhizen aus der *Leccinum-scabrum*-Gruppe. Nova Hedwigia 51(3-4): 381-410. - RAIDL S (1999) *Chamonixia caespitosa* Rolland + *Picea abies* (L.) Karst. Descr Ectomyc 4: 1-6. - SCATTOLIN L, RAIDL S (2006) *Byssoporia terrestris* (DC.) M. J. Larsen & Zak. Descr Ectomyc 9/10: 15-20.- TEDERSOO L, SUVI T, LARSSON E, T, KÕLJALG U (2006) Diversity and community structure of ectomycorrhizal fungi in a wooded meadow. Mycol. Res. IIO(2006) 734-748. - WIEDMER E, SENN-IRLET B, AGERER R (2001) *Alpova diplophloeus* (Zeller & Dodge) Trappe & A. H. Smith + *Alnus viridis* (Chaix) DC. Descr Ectomyc 5: 1-8. - ZAK B (1969) Characterization and classification of mycorrhizae of Douglas fir I. *Pseudotsuga menziesii* + *Poria terrestris* (blue- and orange-staining strains). Can J Bot 47: 1833-1840. - ZAK B, LARSEN MJ (1978) Characterization and classification of mycorrhizae of Douglas fir. III. *Pseudotsuga menziesii* + *Byssoporia (Poria) terrestris* vars. *lilacinorosea, parksii,* and *sublutea.* Can J Bot 56: 1616-1424.

Captions: **Fig. 1.** Habit. – **Fig. 2.** Outer mantle layer with ring-like arranged hyphae and with a gelatinous matrix between the hyphae and a thick gelatinous layer on the surface (shown only below) with bipyramidal to octahedral crystals. – **Fig. 3 – a.** Middle mantle layer, with a slight gelatinous matrix and occasionally ring- to star-like arranged hyphae. – **b.** Inner mantle layer with ring-like arranged hyphae and with a distinct gelatinous matrix. - **Fig. 4.** Thin, probably young rhizomorph composed of loosely packed hyphae; upper and more differentiated portion with crystals and a few peripheral hyphae; the dotted hypha represents the bluish colour after bruising, the asterisk an open anastomosis. – **Fig. 5 – a.** Rhizomorph in a middle developmental stage; central, densely arranged hyphae with irregularly shaped crystals and twisted peripheral hyphae, some of them scarcely ramified. -**b.** Surface of the mantle with staghorn-shaped, cystidia-like hyphae. – **Fig. 6.** Rhizomorph in a middle developmental stage; in optical section of the center, densely arranged hyphae, occasionally with clamps, with twisted peripheral hyphae, some of them scarcely ramified (asterisk). – **Fig. 7.** Plan view of a thick rhizomorph with distinctly twisted hyphae, some remind of a cork-screw; crystals below the peripheral hyphae visible. *All figs. from EDM 55 (in PD).*

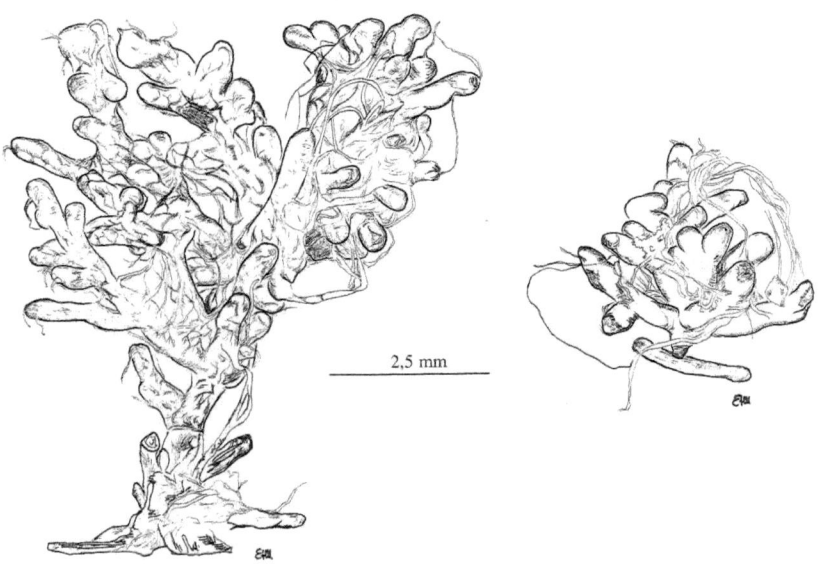

Fig. 1.

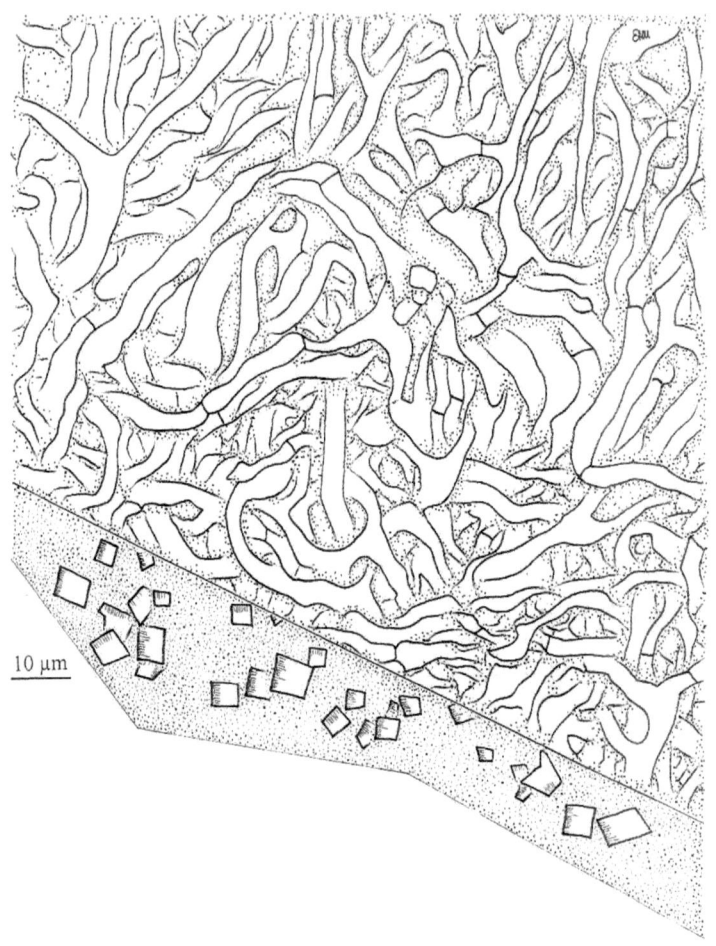

Fig. 2.

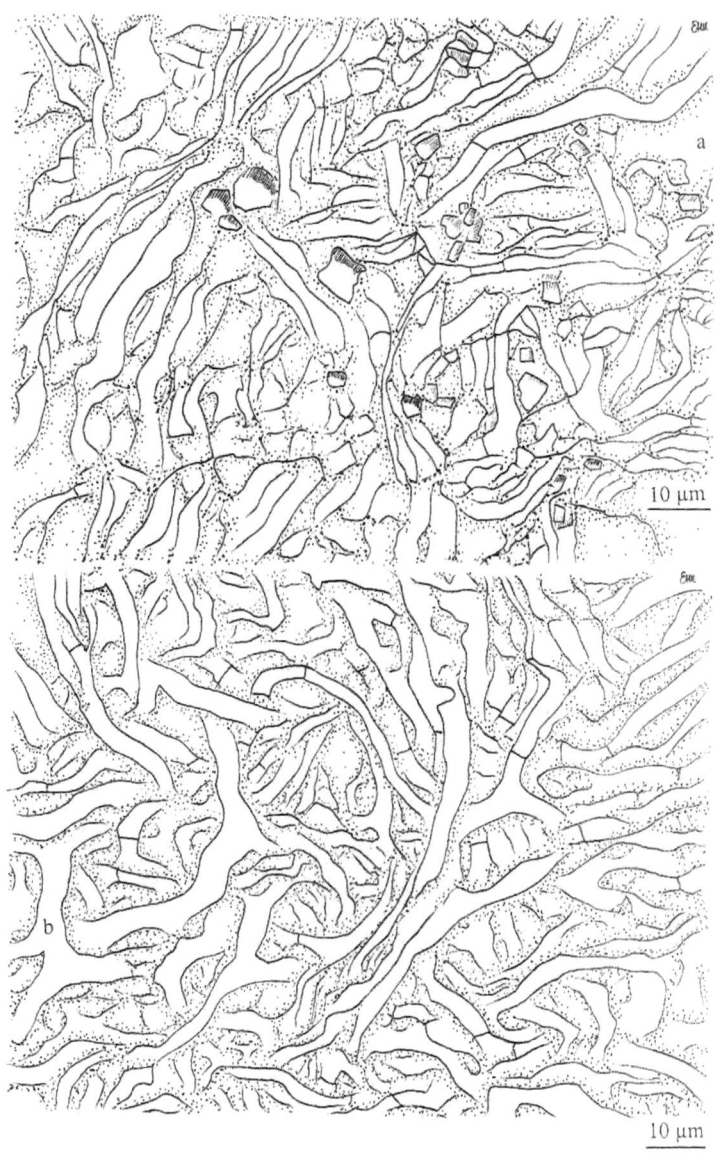

Fig. 3.

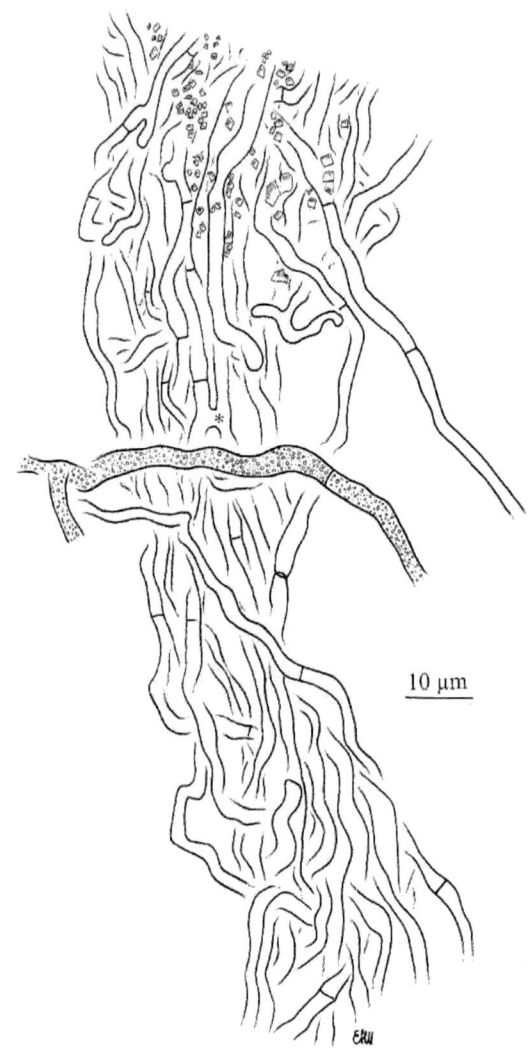

Fig. 4.

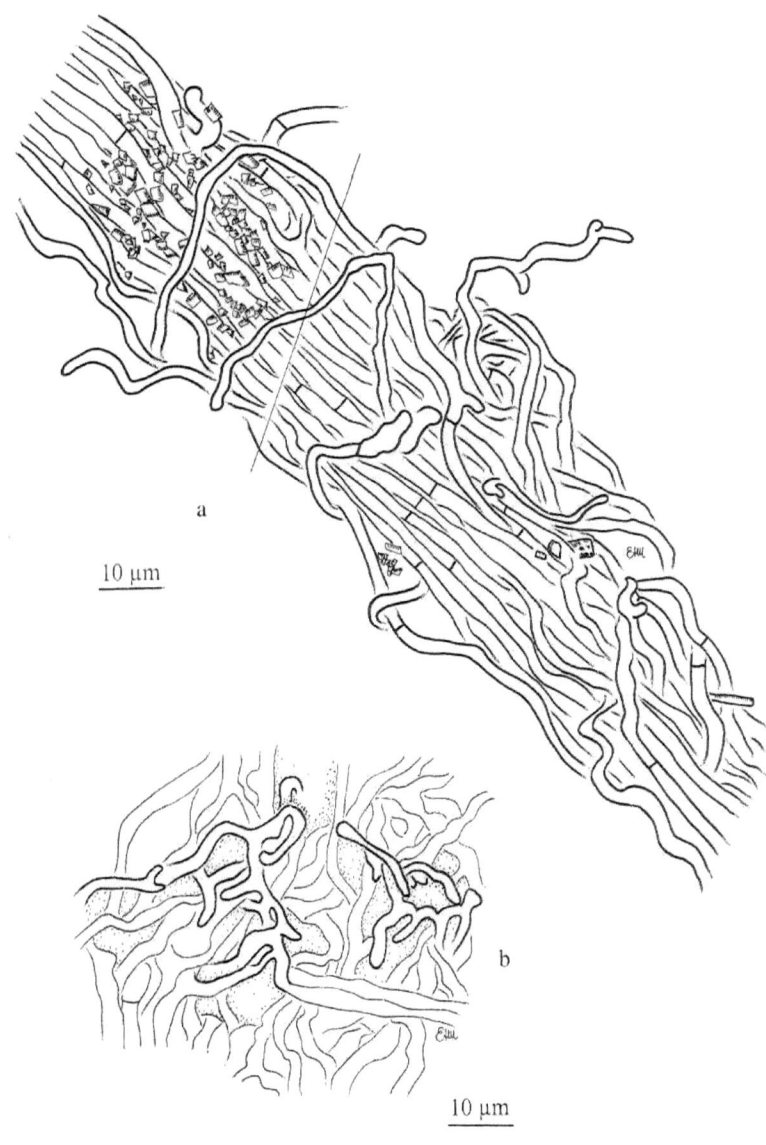

Fig. 5.

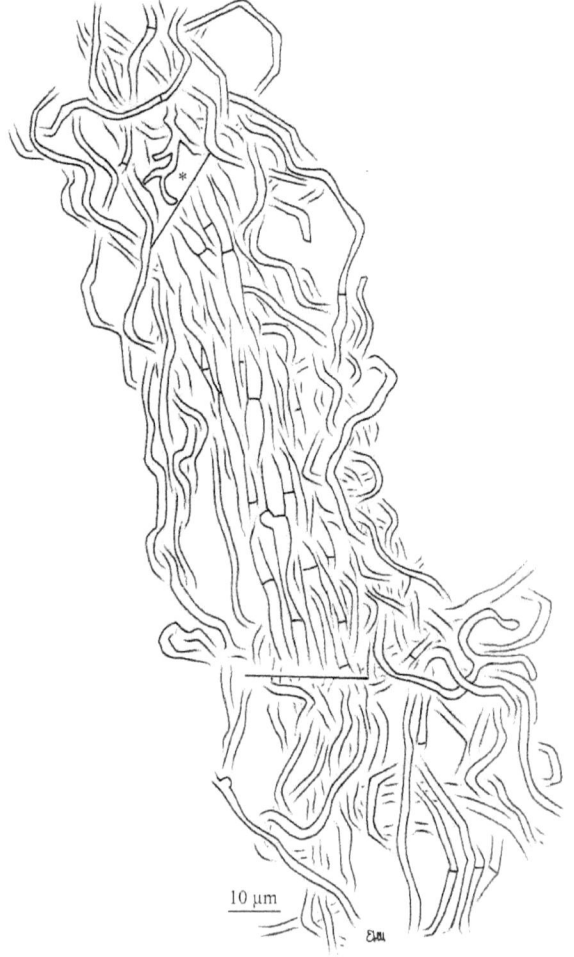

Fig. 6.

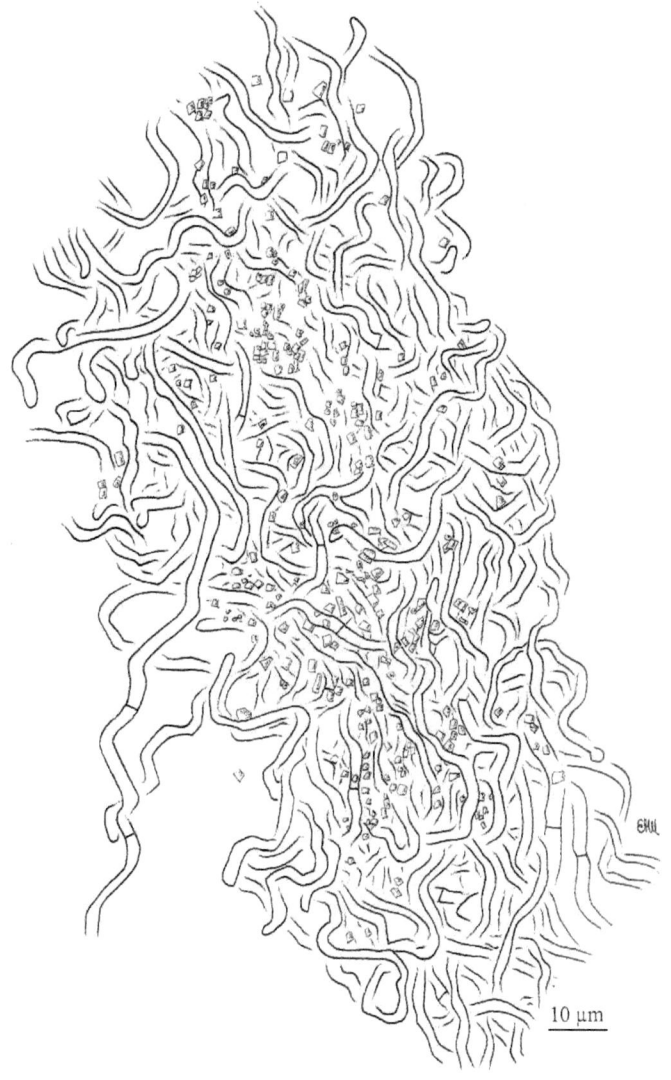

Fig. 7.

CHAPTER 7

"*Fagirhiza stellata*" + *Fagus sylvatica* (L.)

ERIKA DI MARINO[1,2], LUCIO MONTECCHIO[1], REINHARD AGERER[3],

[1] Università degli Studi di Padova Agro-Forestry Systems and Land Use (TeSAF) Department V.le dell'Università, 16 - 35020 Legnaro (PD) – Italy

[2] Centro di Ecologia Alpina - Viote del Monte Bondone (TN), Località Caserme, 2 38060 Garniga Terme (TN)- Italy

[3] Organismic Biology: Mycology, Department Biology and GeoBio-Center[LMU], University of München, Menzinger Str., 67, D-80638, Germany

Short description

The ectomycorrhizae are dark reddish-brown to blackish; older parts dark-brown to black, at maturity, and where air included, with golden tint, monopodial-pyramidal. The mantle is pseudoparenchymatous, with ring- to star-like arrangement on the surface. The middle mantle layer is plectenchymatous, while the inner mantle is transitional between pseudoparenchymatous and plectenchymatous. Rhizomorphs dark brown, surface covered by irregularly shaped, repeatedly ramified, densely entwining thin, rarely septate peripheral hyphae, membranaceously brownish to yellowish, smooth; thinner rhizomorphs lack such hyphae or are only patchily coverered, hyphae with clamps, membranaceously brownish to yellowish. Nodia and conical structures at points of ramification present, slightly differentiated, with infrequent, homogeneously brownish filled hyphae. Cystidia are lacking. The peculiar characteristics, similar to those reported for some ectomycorrhizae of the genus *Tomentella*, are the net of hyphae on the mantle surface, consisting of stars connected by single hyphae or thin hyphal bundles, and the thelephoroid rhizomorphs.

Morphological characters (Fig.1a): *Mycorrhizal systems* abundant, dense and compactly arranged, monopodial-pyramidal, medium distance exploration type of the smooth subtype. - *Main axes* 6 mm long and 0.4 mm diam., straight, up to 2 orders of ramification, with 3-4 side-branches per 10 mm. - *Unramified ends* up to1.5 mm long and 0.2- 0.3 mm diam., not inflated, cylindric, bent to tortuous. - *Surface of unramified ends* dark reddish-brown to blackish; older parts dark-brown to black, very tips reddish brown to blackish , with distinct mantle surface, covered by soil particles; cortical cells not visible and mantle not transparent; mycorrhizal surface loosely cottony, hydrophobic, silvery at patches with slightly golden tint. – *Rhizomorphs* round in cross-section, originating proximally; concolourous to the mantle, brown or dark brown, connection to the mantle distinct; margin of rhizomorphs smooth; frequently ramified at restricted points.- – *Sclerotia not observed.*

Anatomical characters of mantle in plan views: Lacking are cells densely filled with oily droplets, blue granules, needle-like contents, drops of exuded pigment, cell wall projections in pseudoparenchymatous cells. – *Outer mantle layers* (Figs. 2a, 3, 4) pseudoparenchymatous, with ring-like arranged hyphal bundles on the surface (often rather star-like; mantle type P, according to AGERER 1991, 1995, AGERER 1987-2006, AGERER & RAMBOLD 2004-2007); hyphae of the surface net cylindric, not constricted at septa, 1,5-4 µm wide, distance of septa 10-35 µm, with infrequent clamps; angles between hyphal junctions in the net ca. 45° and less up to 120°, membranaceously yellowish to brownish, smooth, cells walls 0.1- 0.5 µm thick; cells of the pseudoparenchymatous outer mantle layer (5)7-10(15) µm diam., (4)10-15(30) µm long, number of cells in a square of 20x20 µm (15)17-20 (32); stars in plan view 15-30 µm in diam. Slight gelatinous matrix present only on the stars' surface. Below the stars, the mantle presents epidermoid cells, (3)5-10 µm long, (4)5-7(10) µm diam. (Figs.5a, 5b) - *Middle mantle layers* (Fig. 1b) pseudoparenchymatous, membranaceously yellowish to brownish, cells 5-7(11) µm diam., (6)15(18) µm long, cell walls 0.2- 0.5 µm wide, smooth, gelatinous matrix is lacking; (13)16-23(29) cellsin a square of 20x20 µm. - *Inner mantle layers* (Fig. 2b) transitional between plectenchymatous and pseudoparenchymatous, membranaceously yellowish to brownish, cells (3)5-13(18) µm diam., (4)8-10(28) µm long, gelatinous matrix is lacking.

Anatomical characters of emanating elements: Lacking are a gelatinous matrix, gelatinized hyphae, ampullate hyphae, drops of secreted pigment, in IC strongly light reflecting crystals, and intrahyphal hyphae. - *Rhizomorphs* (Figs. 6a,b, 7) of type C (6)15-25(35) µm diam, cells homogeneously filled with brownish contents rarely present, 2-3 µm wide, with clamps; rhizomorphs with nodia and conical structures at points of ramification, slightly differentiated, (AGERER 1991, 1995, AGERER 1987-2006, AGERER & RAMBOLD 2004-2007; thelephoroid, AGERER

1999; AGERER & IOSIFIDOU 2004), terminated very rarely by a single hypha; surface covered by irregularly shaped, repeatedly ramified, densely entwining thin, rarely septate, smooth hyphae, hyphae 1-2 μm wide; membranaceously brownish to yellowish, smooth, thinner rhizomorphs (up to 30 μm) such hyphae lacking or covered only patchily; internal central hyphae (1)2-3(4) μm diam., walls 0.2- 0.5 μm wide, membranaceously brownish to yellowish, with clamps. - *Emanating hyphae* not observed. – *Cystidia* not observed. - *Chlamydospores* not observed.

Anatomical characters, longitudinal section: *Mantle* (25)30(45) μm wide, different layers not discernible; hyphae tangentially (3)4-6(7) μm, radially (3)4-6(7) μm. - *Tannin cells* lacking. – *Epidermal cells* tangentially-oval to elliptic or cylindrical and, when oriented-obliquely rectangular, tangentially (20)30-35(45) μm, radially (15)20-25(30) μm; EC_t= ca. 30,5; EC_q = 1,4. *Hartig net in section* paraepidermal, shape of hyphal cells around the epidermal cells beaded, in one row, (2)3-4 μm wide. *Hartig net in plan view* of palmetti type, with 2-5 μm broad lobes.

Colour reaction with different reagents: *Mantle and rhizomorph preparations*: cotton blue: n.r. (no reaction); ethanol 70%: n.r.; FEA: n.r.; guaiac: n.r.. KOH 15%: n.r.; iron (II)sulphate: slightly greyish; sulpho-vanillin: n.r.; KOH 15%: n.r.; lactic acid: n.r., the golden colour is disappearing; Melzer's reagent: slightly greenish, due to the darkness of the mantle, it is difficult to interpret the reaction.

Autofluorescence: *Whole mycorrhiza:* UV 254 nm: lacking; UV 366 nm: lacking. *Mantle in section:* UV-filter 340-380 nm: slightly whitish; blue filter 450-490 nm: slightly yellowish; filter 530-560 nm: slightly reddish. *Rhizomorph:* n.r.

Reference specimen for ectomycorrhiza: Italy, province Trient (Trentino-Alto Adige Region), Val di Non, district Denno (46°14' N; 10° 57' E), beech coppice, 1050-1200 m a.s.l.; myc. isol. E. Di Marino, 20.06.2005, EDM 21 in FEA (in PD); it is supposed that this ECM is a member of the genus *Tomentella* due to similarities to already published descriptions of *Tomentella* ECM. – ***Additional specimens examined***: Italy, province Trient (Trentino-Alto Adige Region), Val di Non, district Denno (46°14' N; 10° 57' E), beech coppices, 1050-1200 m a.s.l.; myc. isol. E. Di Marino, June 2005, EDM 21a in FEA (in PD), October 2005 EDM 21b in FEA (in PD), May/June 2006, EDM 21c in FEA (in PD). – Soil conditions of all collections: mesic or xeric, in mineral layers, pH of the soil 4,8-5,9, N_{tot} 4,6-5,2, C/N 18,5-21, C_{org} 82-108 g/Kg; in organic layers pH 4,8-6, N_{tot} 8,6-26,5, C_{org} 108-382 g/Kg, C/N 16-18.

DNA analysis: Sequencing and alignments were done according to the method applied by Tedersoo et al. (2006). GenBank Accession number EU444548.

Discussion: *Fagirhiza stellata* on *Fagus sylvatica* is similar to some ectomycorrhizae of the genus *Tomentella*, that are still unidentified (DE ROMAN et al. 2005; Agerer & Rambold 2004-2007). Here we compare *F. stellata* with *Quercirhiza stellata*. (DE ROMAN et al. 2002), *Quercirhiza nodulosomorpha* (AZUL et al. 1999), and *Quercirhiza summatriangularis* (AZUL et al. 2006), because they are brownish or blackish, and form an outer mantle with a hyphal net on a pseudoparenchyma.

Quercirhiza. stellata differs from *F. stellata* by its star-like arranged angular cells of the outer mantle layer, and by a plectenchymatous inner mantle layer with ring-like arranged hyphae. The cell walls of the outer mantle layers of *Q. stellata* are often thick and dark at intersection areas of the cells, whereas those of *F. stellata* lack dark intersections as well as thick walls. Furthermore, *F. stellata* differs regarding middle mantle layers from *Q. stellata* in the lack of thick walls and dark intersections between the cells. In contrast to *F. stellata*, emanating hyphae and rhizomorphs are not found in *Q. stellata*.

In contrast to *F. stellata*, *Q. nodulosomorpha* possesses prominent cystidia. The middle mantle layer of *Q. nodulosomorpha* is densely plectenchymatous to almost pseudoparenchymatous with star-like arranged cells, the inner mantle presents a dense plectenchyma, and differs regarding both layers from *F. stellata*. Thelephoroid rhizomorphs, with nodia and conical side-branches occur in both ectomycorrhizae.

A pseudoparenchymatous outer mantle is present in *Q. summatriangularis* as well as in *F. stellata*. But in contrast to *F. stellata* that bears ring-like often rather star-like arranged hyphal bundles on the surface, the mantle of *Q. summatriangularis* is covered by a distinct hyphal net forming triangular rings, with crystals at places. Exclusively the hyphal net of *F. stellata* possesses a slight matrix on its surface. The middle mantle layer of *F. stellata* reveals a pseudoparenchyma, whereas that of *Q. summatriangularis* is plectenchymatous and consists of short, irregularly shaped hyphae. The inner mantle layer of the latter ectomycorrhiza is completely plectenchymatous, whereas that of *F. stellata* forms a transitional type between a plectenchyma and a pseudoparenchyma and shows granular contents in some hyphae. Rhizomorphs could not be found in *Q. summatriangularis*.

The DNA sequence of the ectomycorrhiza presented here agrees best with that of *Tomentella subtestacea* Bourdot & Galzin and *T. bryophila* (Pers.) M. J. Larsen that are both deposited in UNITE (KÕLJALG et al. 2005). The sequence comparison retrieved similarity values of 92% and

91%, respectively. But BLASTn searches in GenBank yielded no unambiguous results. The highest similarity values with the query sequence generally received sequences of *Tomentella* species, the best of them being *T. bryophila* (98%) with a query coverage of only 91%, however.

Acknowledgments: We thank to Dr. S. Raidl, Dr. P. Bodensteiner, Nourou Soulemane Yorou, Dr. L. Beenken, R. Verma, Miss C. Bubenzer-Hange, Mr. E. Marksteiner for discussions and technical help, the "A. Gini Foundation" of Padua University for the partial financial support, and the Centre of Alpine Ecology of Monte Bondone (TN-Italy) for soil analysis.

References: AGERER R (1987-2006) Colour atlas of ectomycorrhizae. 1th -13th del. Einhorn, Schwäbisch Gmünd, Germany. - AGERER R (1991) Characterization of ectomycorrhiza. In Norris JR, Read DA, Varma AK (eds) Techniques for the study of mycorrhiza. Methods in Microbiology, vol 23, Academic Press, London et al., pp 25-73. - AGERER R (1995) Anatomical characteristics of identified ectomycorrhizas: an attempt towards a natural classification. In Varma K, Hock B (eds) mycorrhiza: structure, function, molecular biology and biotechnology. Springer, Berlin, Heidelberg, New York, pp 685-734. - AGERER R (1999) Never change a functionally successful principle: the evolution of Boletales s. l. (Hymenomycetes, Basidiomycota) as seen from below-ground features. Sendtnera 6: 5-91. - AGERER R (2006) Fungal relationship and structural patterns of their ectomycorrhizae. Mycological Progress 5(2): 67-107. - AGERER R, IOSIFIDOU P (2004) Rhizomorph structures of Hymenomycetes: a possibility to test DNA-based phylogenetic hypotheses? In Agerer R, Piepenbring M, Blanz P (eds) Frontiers in Basidiomycote Mycology, IHW-Verlag, Eching, pp 249-302. - AGERER R, RAMBOLD G (2004-2007) [first posted on 2004-06-01; most recent update: 2007-05-02]. DEEMY – An Information System for Characterization and Determination of Ectomycorrhizae. www.deemy.de – München, Germany. - AZUL A M, AGERER R, FREITAS H (1999) "*Quercirhiza nodulosomorpha*" + *Quercus suber* L..Descr Ectomyc 4:103-108 (1999). - AZUL A M, AGERER R, PAZ MARTIN M (2006) "*Quercirhiza summatriangularis*" + *Quercus suber* L..Descr Ectomyc 9/10:111-114 (2006). - DE ROMAN M, AGERER R, DE MIGUEL A (2002):"*Quercirhiza summatriangularis*" + *Quercus ilex* L. subsp. *ballota* (Desf.) Samp. Descr Ectomyc 6: 19-24 (2002). - DE ROMAN M, CLAVERIA V, DE MIGUEL AM (2005) A revision of the description of ectomycorrhiza published since 1961. Mycol. Res. 109(10):1063-1104. - KÕLJALG U, LARSSON K-H, ABARENKOV K, NILSSON RH, ALEXANDER IJ, EBERHARDT U, ERLAND S, HOILAND K, KJOLLER R, LARSSON E, PENNANEN T, SEN R, TAYLOR AFS, TEDERSOO L, VRALSTAD T, URSING BM. (2005) UNITE: a database providing web-based methods for the molecular identification of ectomycorrhizal fungi. New Phytol 166: 1063-1068. - TEDERSOO L, SUVI T, LARSSON E, KÕLJALG

U (2006) Diversity and community structure of ectomycorrhizal fungi in a wooded meadow. Mycol. Res. 110 734-748.

Captions: **Figs. 1** – **a.** Habit. - **b.** Pseudoparenchymatous middle mantle layer. - **Fig. 2-a.** Plan view of outer mantle, at one place with slightly larger cells surrounding smaller ones (such structures occur preferentially below stars of the surface net). - **b.** Inner mantle layer: transitional type between pseudoparenchymatous and plectenchymatous, showing some large cells (often occurring below stars of the surface net). - **Fig. 3.** Plan view of the star-like arranged hyphal surface net with bundles of hyphae connecting the stars, stars sometimes connected only by solitary hyphae. - **Fig. 4-a.** Open anastomosis with a short bridge, and a hypha showing a globularly inflated cell. - **b.** Hyphal surface net connected to the pseudoparenchyma of the outer mantle layer. **Fig. 5-a.** Plan view of mycorrhizal surface with a star showing a slight matrix. - **b.** Epidermoid to irregularly shaped cells beneath a star of the surface net (the same position as 'a'). - **Fig. 6 - a.** Rhizomorph with one hypha homogeneously filled by brownish contents -**b.** Thinner rhizomorph with nodium and conical structure, together with infrequently septate peripheral hyphae. - **Fig. 7.** Thick rhizomorph partially covered by thin hyphae (above), the optical section (below) shows a hypha with clamp. *All figs. From EDM 21a (in PD).*

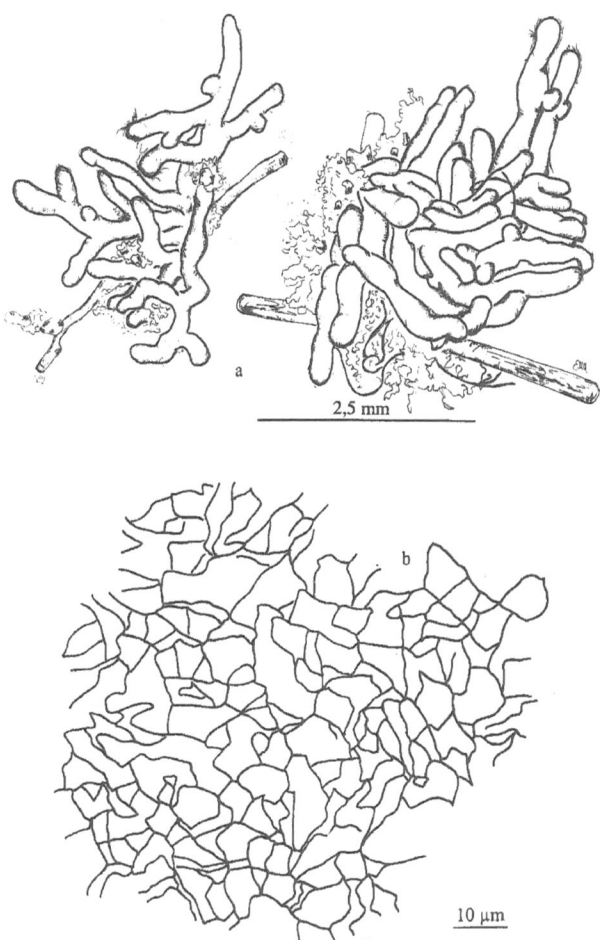

Figs. 1a, 1b.

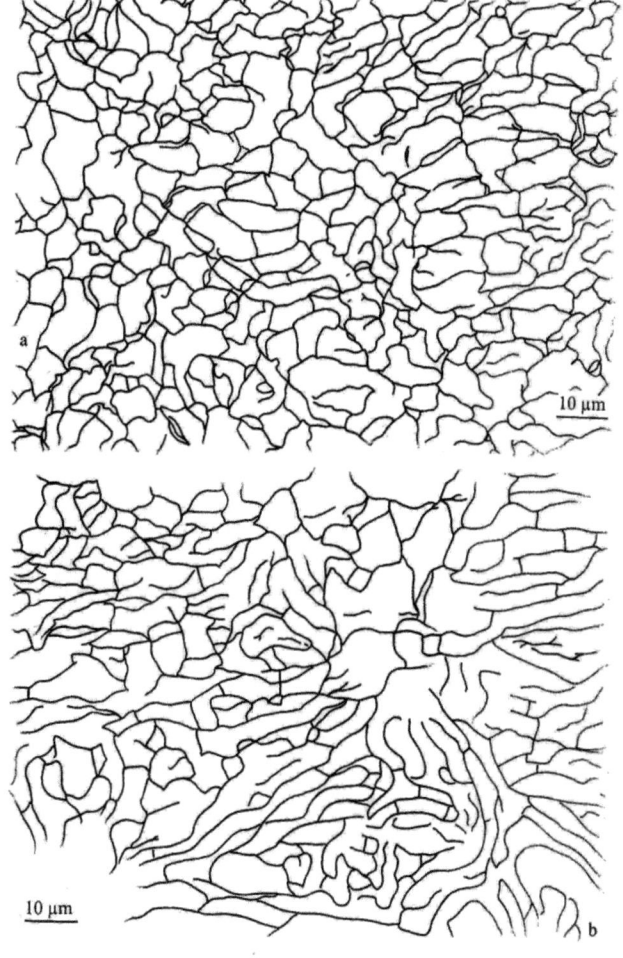

Figs. 2a, 2b.

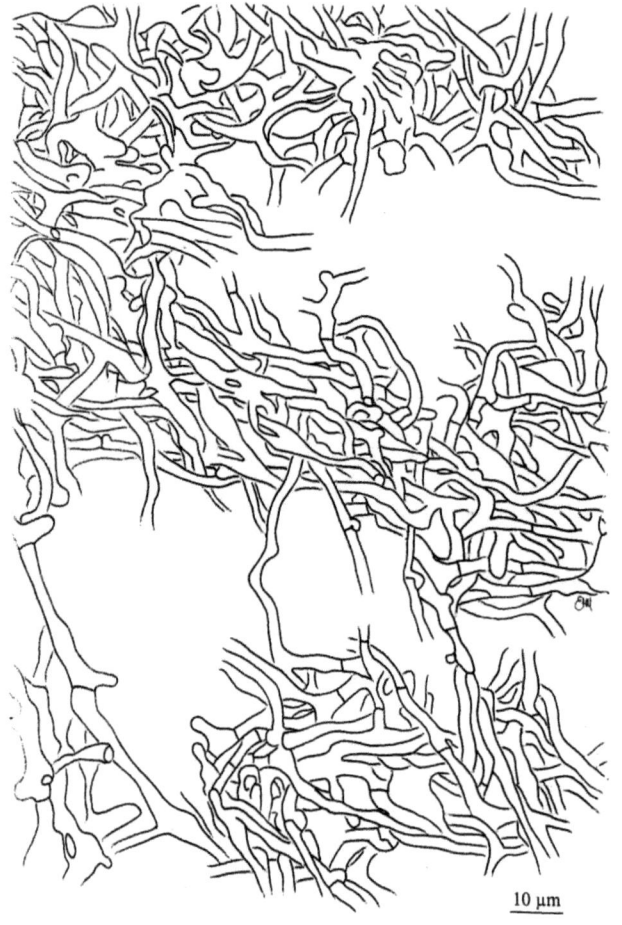

Fig. 3.

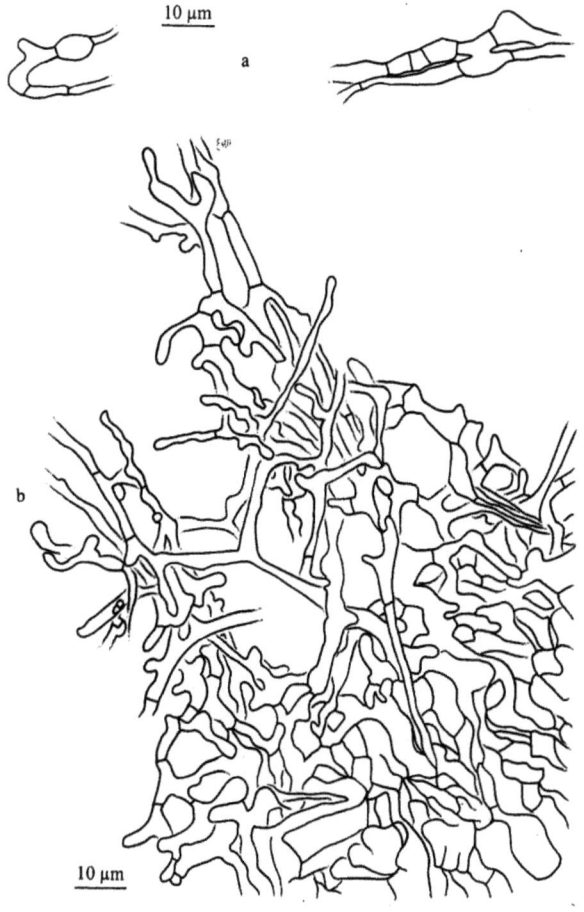

Figs. 4a, 4b.

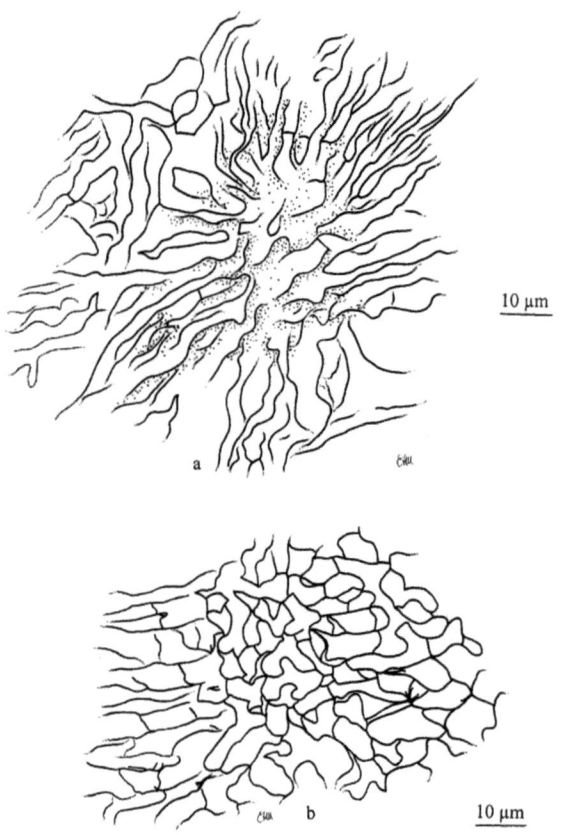

Figs. 5a, 5b.

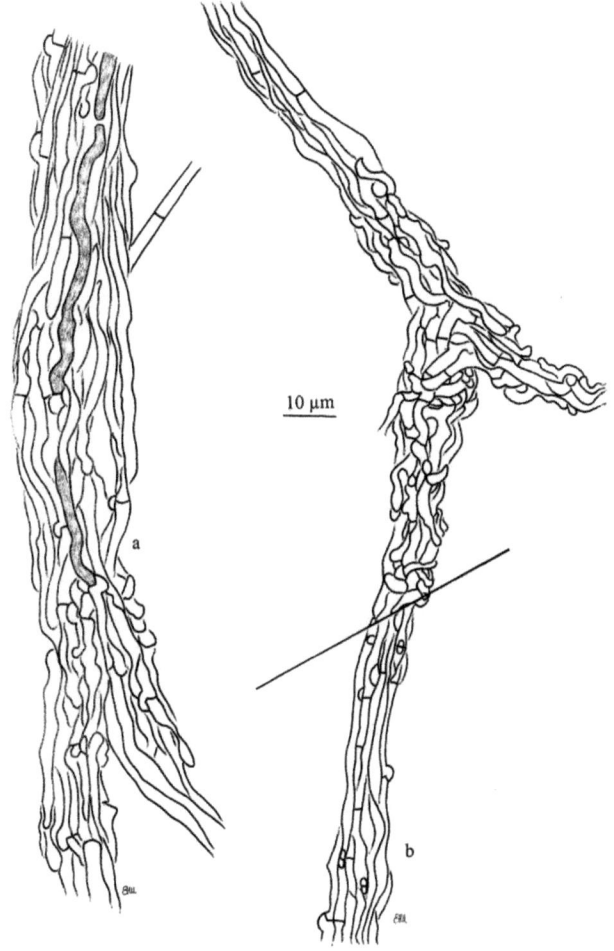

Fig. 6.

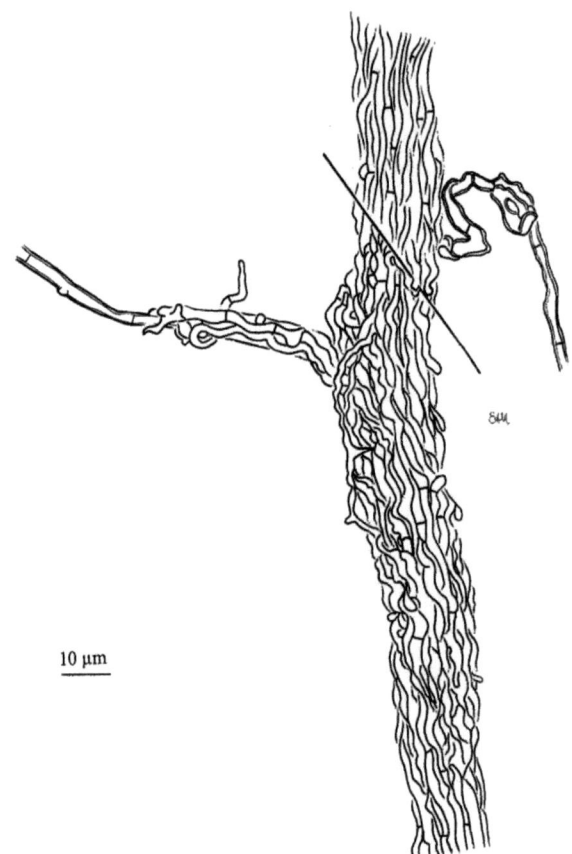

Fig. 7.

CHAPTER 8

Sistotrema is a genus with ectomycorrhizal species – confirmation of what sequence studies already suggested

Submitted the 11th Jenuary 2008 to Mycological Research

Erika Di Marino[1], Linda Scattolin[2], Philomena Bodensteiner[3], and Reinhard Agerer[3)*]

[1] *Università degli Studi di Padova, Agro-Forestry Systems and Land Use (TeSAF) Department, V.le dell'Università, 16 - 35020 Legnaro (PD), Italy*

[2] *Centro di Ecologia Alpina - Viote del Monte Bondone; I 38070 Sopramonte (TN), Italy*

[3] *Organismic Biology: Mycology, Department Biology and GeoBio-Center^{LMU}, University of Munich, Menzinger Str. 67, 80638, Germany*

*) Corresponding author

Abstract

The ectomycorrhizal status of Sistotrema muscicola is shown the first time unequivocally, although already before sistotremoid DNA had been extracted from ectomycorrhizae (ECM). The ECM are irregularly monopodial-pyramidal, whitish ochre to yellow ochre, and woolly. When older the ectomycorrhizae become more greyish and silvery at some patches. Diagnostic anatomical characteristics are irregularly inflated emanating hyphae and rhizomorph hyphae, ampullately inflated clamps, and the occurrence of yellow drops within the hyphae. The plectenchymatous mantle shows ring-like arranged hyphae, and a slightly gelatinous matrix. The ECM of *S. muscicola* are compared to those of other species that form distinctly ampullate hyphae in rhizomorphs, too. The anatomically most similar ECM to those of Sistotrema muscicola are those of *Hydnum repandum*.

Introduction

Results of different DNA-based phylogenetic studies recently confirmed that the genus *Sistotrema* is a member of the cantharelloid clade (Binder *et al.*, 2005; Hibbett & Binder 2002; Larsson *et al.*, 2004; Moncalvo *et al.*, 2006). According to them the cantharelloid clade comprises the genera *Botryobasidium, Clavulina, Haplotrichium, Hydnum, Membranomyces,* and *Sistotrema* (Larsson *et al.*, 2004), *Botryobasidium, Cantharellus, Ceratobasidium, Hydnum,* and *Sistotrema* (Binder *et* *al.,*2005), *Botryobasidium, Cantharellus, Clavulina, Craterellus, Hydnum, Multiclavula, Sistotrema, Tulasnella,* and *Uthatobasidium (*Hibbett & Binder, 2002) or *Botryobasidium, Cantharellus, Ceratobasidium, Clavulina, Craterellus, Hydnum, Membranomyces, Multiclavula,* and *Sistotrema* (Moncalvo *et al.,*2006).

In case of some of these genera the ectomycorrhizal status of representative species has already been proven by identification and anatomical characterization of their ectomycorrhizae. This applies to *Hydnum repandum* L. (Agerer *et al.*, 1996), *Cantharellus cibarius* Fr. (Danell 1994; Froideveaux 1975; Mleczko 2004a; Moore *et al.*, 1989; Zak 1973), *Cantharellus formosus* Corner (Countess and Goodman 2000), and *Craterellus tubaeformis* (Bull.) Quél. (Fransson 2004; Mleczko 2004b; Trappe *et al.*, 2000). For *Clavulina crista*ta (Holmsk.) J. Schröt. the obtained DNA sequences suggest an ectomycorrhizal status of at least that member of the genus (Buée *et al.*, 2005; 2007; Dickie *et al.*, 2002; Ogawa 1984; Tedersoo *et al.*, 2003). The remaining genera placed in the cantharelloid clade are still waiting for an unequivocal proof that they contain ectomycorrhizal species.

Based on the comparison of DNA obtained from fruitbodies of *Sistotrema muscicola* (Pers.) S. Lundell and *S. alboluteum* (Bourdot & Galzin) Bondartsev & Singer and that from ectomycorrhizae collected below these basidiomata Nilsson *et al.* (2006) reported on the ectomycorrhizal status of these *Sistotrema* species. In this context provided accompanying colour pictures of the putative *S. muscicola* mycorrhiza show habit and mantle surface of a dark brown ECM with superficial colourless mycelium and an inflated portion at a hyphal septum, which is regarded as a feature typical for Sistotrema mycelia.

In the present contribution we now provide unequivocal evidence that *Sistotrema muscicola* is an ectomycorrhizal species.

Material and methods

The characterization of ECM is comprehensively described in Agerer (1991). Fresh material was studied regarding morphology, colour of hyphae, and chemical reactions; material fixed in FEA (see Agerer 1991) was used for anatomical studies by the aid of a ZEISS Axioskop with Normarski's Interference Contrast connected to a drawing mirror. All drawings were made at a magnification of 2000×, subsequently transferred to transparent paper, and finally reduced in mag-nification.

Identification was possible by the comparison of newly generated nuclear rDNA ITS sequences bounded by primers ITS1-F and ITS4 (Gardes & Bruns, 1993; Vilgalys & Hester, 1990; White *et al.,* 1990; for primer sequences also see http://pmb.berkeley.edu/~bruns/tour/primers.html and http://www.biology.duke.edu/fungi/mycolab/primers.htm) obtained from the mycorrhizal root tips and from the fruitbody that had been collected in close vicinity. For the determination of the fruitbodies Eriksson *et al.*(1984) and Jülich (1984) were used. The applied methods of DNA extraction,

PCR, and sequencing follow Tedersoo *et al.* (2006). GenBank accession numbers of the generated sequences of *Sistotrema muscicola* are 1052862 (fruitbody) and 1052863 (ECM). Reference specimens of the mycor-rhizae and the fruitbodies are deposited in M (see Holmgren *et al.*, 1990). The collection data of the characterized material are as follows: Italy, province of Parma (Emilia-Romagna Region), Vighini, south of Borgotaro, northern exposed slope, *Castanea sativa* forest, ca. 820 m NN, leg. et det. R. Agerer, 30. 10. 2006 (fruitbody RA 14583, ectomycorrhizae RA 14583a).

Results
Description of *Sistotrema muscicola* + *Castanea sativa* **L. ectomycorrhizae**

Morphological characters (Figs. 1a, b): Mycorrhizal systems whitish ochre, the older parts greyish and silvery at places; with 1–2 orders of ramification, 8–9 side branches per cm; medium-distance stringy exploration type, hydrophobic. – Main axes up to 3 mm long and (0.3)0.5–0.4(0.6) mm diam. - Unramified ends white ochre, up to (0.4)0.5–1.3 mm long and 0.3–0.5 mm diam., bent and tortuous, not inflated; older tips greyish. – Surface of unramified ends woolly, very tip not smooth, with a lot of emanating hyphae forming fans. – Rhizomorphs very abundant in older mycorrhizal systems, whitish, woolly, originating also at the very tips of mycorrhizal systems; flat in cross-section, whitish; not distinctly connected to the mantle. – Sclerotia not observed.
Anatomical characters of mantle in plan views (Figs. 2-5): Mantle surface hyphae with many yel-low oily droplets, with a slightly gelatinous matrix between hyphae, with clamps. – Outer mantle layers (Fig. 2) plectenchymatous, with ring-like pattern and slightly gelatinous matrix (mantle type A/C and at places B/C, according to Agerer 1991, Agerer 1987-2006, Agerer and Rambold 2004-2007); hyphae cylindric, not constricted at septa, irregularly shaped; simple septate and with infre-quent clamps; angles between hyphal junctions ca. 45-90°, membranaceously yellowish, smooth, contents with droplets, cell walls thin; cells (2)3–4(5) µm diam. – Middle mantle layers (Fig. 3) plectenchymatous, slightly ring-like, with simple septa, membranaceously yellowish, containing fewer droplets than hyphae of the outer mantle; hyphae cylindric, not constricted at septa, irregu-larly shaped, cell walls thin, smooth, cells (3)4–5(7) µm diam. – Inner mantle layers (Fig. 4) plec-tenchymatous, with occasionally ring-like arranged hyphae and slightly gelatinous matrix, hyphae cylindric, not constricted at septa, membranaceously yellowish, irregularly shaped with infrequent simple septa, hyphae with some internal droplets; cell walls thin, smooth, cells (2)3–4(5) µm diam. – Very tip (Fig. 5) with many clamps and droplets within the hyphae, with slightly gelatinous ma-trix; cells of the outer mantle (3)4–5(6) µm diam., cell walls thin; cells of the inner mantle (1)2–3(5) µm wide.

Anatomical characters of emanating elements (Figs. 6-8): Rhizomorphs with slightly gelatinous matrix, of type C (Agerer 1991, Agerer 1987-2006, Agerer & Rambold 2004-2007, Agerer and Iosifidou 2004), also called ramarioid (Agerer 1999), (20)45–75(120) µm diam., with open anastomoses or closed by a clamp; clamps very frequently strongly inflated; middle portions of hyphae often inflated, there (6)7–8(9) µm diam., not inflated portions 4–5 µm diam.; thinner rhizomorphs undifferentiated, 13–15 µm diam. (Fig. 8), with clamps but without irregularly inflated hyphal portions, cells of the peripheral hyphae occasionally irregularly shaped; cell walls thin. – Emanating hyphae (Fig. 7) frequent, few of them with irregularly branched, irregularly inflated hyphal portions similar to those on the surface of thicker rhizomorphs, with simple septa, and with ampullately inflated clamps, inflated hyphal portions (6)7–8(9) µm diam., not inflated portions 4–5 µm diam. – Chlamydospores not observed.

Anatomical characters, longitudinal section: Tannin cells lacking. – Mantle not compact, thin, 15–20 µm wide, very tips with 12.5–15 µm wide mantle; different layers not discernible; hyphae tangentially (1.2)2.5–3.5(5) µm and radially (1.5)2–2.5 µm. – Hyphal cells around epidermal cells roundish, in one row. – Epidermal cells radially-oval to elliptic, oriented obliquely, tangentially (5)8–11(15) µm, radially (25)30–40(45) µm; EC_t = 8.6 µm , EC_q = 0.25. – Hartig net in section paraepidermal. – Hartig net in plan view of palmetti-type, lobes without septa, lobes (1.2)2.5–3 µm broad.

Colour reaction with different reagents: Mantle and rhizomorph preparations: cotton blue: slightly bluish; FEA: the oily droplets not visible; lactic acid: n.r.; Melzer's reagent: n.r.

Autofluorescence: Rhizomorph in section: UV-filter 340-380 nm: n.r; blue filter 450-490 nm: n.r.; green-filter 530-560 nm: n.r.

DNA sequence data:

PCR products of the targeted ITS rDNA (complete sequence of internal transcribed spacer 1, 5.8S ribosomal RNA region, and internal transcribed spacer 2, flanked by partial sequences of 18S and 28S ribosomal RNA genes, respectively) obtained from fruitbody and ECM material of Sistotrema muscicola have a size of 563 bp (fruitbody; RA 14583) and 553 bp (ECM; RA 14583a). Both are nearly identical with a sequence identity of >99% corresponding to three differing bases, one in the 5.8S region, two in the internal spacer region 2, which may be due to sequencing and/or sequence editing errors.

BlastN searches were performed both in GenBank and UNITE (see Kõljalg et al., 2005) using the newly generated sequences of the *Sistotrema muscicola* fruitbody and ECM as query. The thereby retrieved best matches had a maximum sequence identity of at the most 92%. With values of 92%, 89%, 88% compared to the S. muscicola fruitbody sequence and 91%, 88%, 87 % compared to the

S. muscicola ECM sequence, respectively, the three most similar sequences (AY702760, AB251813, and AB211250) were generated from samples of indetermined, ectomycorrhizal fungi. So far published, completely identical ITS sequences (AJ606040, AJ606041) obtained from fruitbodies and ECM determined as *S. muscicola* in Nilsson *et al.* (2006) retrieved identity values of 87% or 86%, respectively, compared with the fruitbody and ECM sequences of the here characterized S. muscicola specimen.

Discussion

The relatively moderate identity values of the ITS sequences obtained from fruitbody material of the here characterized S. muscicola specimen (RA14583) and the previously published *S. muscicola* collection (Nilsson *et al.*, 2006) rise questions regarding the currently applied species concept of and putative areal and/or host related differences within this taxon, especially as they correspond to slight variations of morphological and anatomical characters. The sequence data of the latter were obtained from Finn-ish material on *Alnus*, whereas our collection was found in northern Italy in a *Castanea sativa* stand. *Sistotrema muscicola* as characterized by Eriksson *et al.* (1984) forms basidia with consistently six sterigmata, those of our collection possess predominantly eight of them. Although the hymenium is described as "hydnoid-irpicoid with teeth 1-2 mm long, cylindrical, conical or more or less flattened, or poroid, at first reticulate with thin, fimbriate or more or less lacerate dissepiments" (Eriksson *et al.*, 1984), our collection is definitely reticulate or shallowly poroid (Figs. 9, 10). Spore characteristics and other anatomical features, on the other hand, fit well those published by Eriksson *et al.* (1984). Putative taxonomic consequences of these observed intraspecific differences require more detailed studies using additional data, however.

The ECM of *Sistotrema muscicola* belong to a group of species that form ampullate inflations within rhizomorphs mostly below a simple hyphal septum or with the incorporation of clamps. This characteristic is most typically presented also by members of *Gomphales*, *Clavariadelphus pistillaris* (L.) Donk (Iosifidou & Raidl, 2006), *Gautieria inapire* Palfner & E. Horak (Agerer 1999; Palfner 2001; Palfner & Horak, 2001), *Gomphus clavatus* (Pers.) Gray (Agerer *et al.*, 1998), *Ramaria aurea* (Schaeff.) Quél. (Agerer 1996a), *R. flavo-saponarea* R.H. Petersen (Scattolin & Raidl, 2006), *R. largentii* Marr & D.E. Stuntz (Agerer 1996b), *R. spinulosa* (Pers.) Quél. (Agerer 1996c), R. subbotrytis (Coker) Corner (Agerer 1996d), Geastrales, Geastrum fimbriatum Fr. (Agerer & Beenken, 1998), Hysterangiales, Hysterangium stoloniferum Tul. & C. Tul (Raidl & Agerer, 1998), and Hydnaceae, Hydnum repandum L. (Agerer *et al.*, 1996).

Gomphales, Geastrales and *Hysterangiales* form a well defined anatomy-based relationship that under inclusion of the saprotrophic orders *Phallales* and *Gastrosporiales* has been established as a new superorder *Gomphanae* in Agerer (1999) as well as Agerer and Iosifidou (2004). Based on results of molecular analyses this group including the *Phallales* was later defined as subclass Phallomycetidae within the *Agaricomycetes* by Hosaka *et al.* (2006). Hydnum repandum is not related to the *Phallomycetidae* as this species belongs to the cantharelloid clade (Hibbett 2006; Moncalvo et al. 2006) or *Cantharellales* (Hibbett 2006), respectively.

Sistotrema and *Hydnum* with ampullate hyphae in rhizomorphs are both members of the cantharelloid clade. Ectomycorrhizal rhizomorphs of *Cantharellus cibarius* (Mleczko 2004a) also form slightly inflated clamps, and *Craterellus lutescens*, a species that lacks rhizomorphs (Mleczko 2004b), shows those on the ECM mantle surface. Ampullately inflated hyphae are also reported for *Clavulina* spp. (Breitenbach & Kränzlin, 1986). Although such inflations have been evolved independently at least three times as they occur in rhizomorphs of species in the trechisporoid clade (Agerer & Iosifidou, 2004; Nilsson *et al.*, 2006) and in *Gastrosporium*, too (Iosifidou & Agerer, 2002), this special anatomical feature can, at least according to the present state of knowledge, be considered as a common character of the ectomycorrhizal members of the cantharelloid clade.

Apart from the typical ampullate inflations the ECM of *Sistotrema muscicola* are characterized by their woolly surface, the whitish ochre to yellow ochre colour that becomes more greyish and silvery at some patches, and by a plectenchymatous mantle with ring-like patterns together with a slightly gelatinous matrix and yellowish droplets within the hyphae. The occurrence of yellowish droplets is a feature that also occurs in ECM of *Cantharellus cibarius* (Mleczko 2004a).

Most frequently the colour of the substrate mycelium of fruitbodies – no matter if growing superficially or within the substrate – corresponds to the colour of the ECM. Generally, dark brown ECM are not formed by species that possess colourless or slightly yellowish hyphae, and, vice versa, all hitherto identified and described dark brown ECM are the symbiotic organs of fungi with brown hyphae (Agerer 1987-2006, Agerer 2006, Agerer 2007, Agerer, unpubl., Agerer & Ram-bold 2004-2007). Therefore, the picture of ECM that has been attributed to Sistotrema muscicola by Nilsson *et al.* (2006) is very likely the result of a misidentifcation. Quite frequently ECM of different fungal species can be overgrown by mycelium of a diversity of fruitbodies. This is obviously the case in the microscopical picture in Nilsson *et al.* (2006) that shows a typical *Sistotrema* hypha laying loosely on the mantle surface of a dark brown ECM. It seems comprehensible that DNA may more easily be extracted from thin-walled, living hyphae that envelope a foreign dark brown ECM than from the underlying ECM composed of thick-walled hyphae with often degenerated cytoplasm.

The current study, therefore, presents the first unequivocal report that the genus Sistotrema contains at least one ectomycorrhizal species, although molecular evidence suggested this ecologi-cal status before (Nilsson *et al.*, 2006).

Sistotrema muscicola ECM clearly differ from all hitherto described ECM of the genera *Clava-*

riadelphus, *Gautieria*, *Geastrum*, *Gomphus*, *Hysterangium*, and *Ramaria* (Agerer 2006, Agerer & Rambold 2004-2007) by the lack of oleoacanthocystidia and/or oleoacanthohyphae and by thin-walled cells with yellowish contents. Much more difficult is the distinction from ECM of *Hydnum repandum*, as the latter forms yellowish drops within the hyphae, too. Unlike Sistotrema muscicola H. repandum shows rough hyphae and orange irregularly shaped or crystal-like encrustations on the hyphal surface, however. Additionally, rhizomorphs of H. repandum provide on their peripheral hyphae sometimes very thin branches that are unknown in *Sistotrema muscicola*.

Acknowledgements

We thank the "A. Gini Foundation" of the University of Padua (Italy) and the Centre of Alpin Ecology of Monte Bondone (TN-Italy) for the financial support of the presented study.

References

AGERER R (1987-2006). Colour atlas of ectomycorrhizae. 1st –13th del. Einhorn, Schwäbisch Gmünd, Germany

AGERER R (1991). Characterization of ectomycorrhiza. In: Norris JR, Read DA, Varma AK (eds) Techniques for the study of mycorrhiza. Methods in Microbiology, vol 23, Academic Press, London et al., pp 25–73.

AGERER R (1996a). *Ramaria aurea* (Schaeff.: Fr.) Quél. + *Fagus sylvatica* L. *Descr Ectomyc* 1:107–112.

AGERER R (1996b). *Ramaria largentii* Marr & D. E. Stuntz + *Picea abies* (L.) H. Karst. *Descr Ectomyc* 1:113–118.

AGERER R (1996c). *Ramaria spinulosa* (Fr.) Quel. + *Fagus sylvatica*. *Descr Ectomyc* 1:119–124

AGERER R (1996d). *Ramaria subbotrytis* (Coker) Corner + *Quercus robur* L. *Descr Ectomyc* 1:125–130.

AGERER R (1999). Never change a functionally successful principle: the evolution of *Boletales* s. l. (*Hymenomycetes*, *Basidiomycota*) as seen from below-ground features. *Sendtnera* 6:5–91

AGERER R (2006). Fungal relationships and structural identity of their ectomycorrhizae. *Mycol Progress* 5:67–107.

AGERER R (2007). Diversität der Ektomykorrhizen im unter- und oberirdischen Vergleich: die Explorationstypen. *Z Mykol* 73:61–88.

AGERER R, BEENKEN L (1998). *Geastrum fimbriatum* Fr. + *Fagus sylvatica* L. *Descr Ectomyc* 3:13–18.

AGERER R, BEENKEN L, CHRISTAN J (1998). *Gomphus clavatus* (Pers.: Fr.) S.F. Gray + *Picea abies* (L.) Karst. *Descr Ectomyc* 3:25–29.

AGERER R, IOSIFIDOU P (2004). Rhizomorph structure of Hymenomycetes: a possibility to test DNA-based phylogenetic hypotheses? In: Agerer R, Piepenbring M, Blanz P (eds) Frontiers in Basi-diomycote Mycology, IHW-Verlag, Eching, pp 249–302.

AGERER R, KRAIGHER H, JAVORNIK B (1996). Identification of ectomycorrhizae of *Hydnum rufescens* on Norway spruce and the variability of the ITS region of H. rufescens and *H. repandum* (Basidio-mycetes). *Nova Hedwigia* 63:183–194.

AGERER R, RAMBOLD G (2004-2007) [first posted on 2004-06-01; most recent update: 2007-05-02]. DEEMY – An Information System for Characterization and Determination of Ectomy-corrhizae. www.deemy.de – München, Germany

BINDER M, HIBBETT DS, LARSSON K-H, LARSSON E, LANGER E, LANGER G (2005). The phylogenetic dis-tribution of resupinate forms across the major clades of mushroom-forming fungi (Homobasi-diomycetes). *Systematics and Biodiversity* 3:113–157.

BREITENBACH J, KRÄNZLIN F (1986). Pilze der Schweiz 2. Nichtblätterpilze. Luzern

BUÉE M, COURTY PE, MIGNOT D, GARBAYE J (2007). Soil niche effect on species diversity and cata-bolic activities in an ectomycorrhizal fungal community. *Soil Biol Biochem* 39(8):1947–1955.

BUÉE M, VAIRELLES D, GARBAYE J (2005). Year-round monitoring of diversity and potential metabolic activity of the ectomycorrhizal community in a beech (*Fagus silvatica*) forest subjected to two thinning regimes. *Mycorrhiza* 15:235–245.

COUNTESS RE, GOODMAN DM (2000). *Cantharellus formosus* Corner + *Tsuga heterophylla* (Raf.) Sarg. In: Goodman DM, Durall DM, Trofymow JA, Berch SM (eds) Concise Descriptions of North American Ectomycorrhizae. Mycologue Publications, and Canada-B. C. Forest Resource Development Agreement, Canadian Forest Service, Victoria, B. C., pp CDE21.1-CDE21.4

DANELL E (1994). Formation and growth of the ectomycorrhiza of *Cantharellus cibarius*. *Mycorrhiza* 5:89–97.

DICKIE IA, XU B, KOIDE RT (2002). Vertical niche differentiation of ectomycorrhizal hyphae in soil as shown by T-RFLP analysis. *New Phytol* 156:527–535.

ERIKSSON J, HJORTSTAM K, RYVARDEN L (1984). The Corticiaceae of North Europe 7. *Schizopora – Suillosporium*. Oslo

FRANSSON PMA (2004). *Craterellus tubaeformis* (Fr.) Quél. (syn. *Cantharellus tubaeformis* Fr.: Fr.) + *Quercus robur* L. *Descr Ectomyc* 7/8:37–43.

FROIDEVEAUX L (1975). Identification of some douglas fir mycorrhizae. *Eur J For Pathol*

5:212–216

GARDES M, BRUNS TD (1993). ITS primers with enhanced specificity for basidiomycetes – applica-tions to the identification of mycorrhizae and rusts. *Mol Ecol* 2:113–118.

HIBBETT D (2006). A phylogenetic overview of the Agaricomycotina. Mycologia 98:917–925.

HIBBETT DS, BINDER M (2002). Evolution of complex fruiting-body morphologies in homobasidio-mycetes. *Proc R Soc Lond B* 269:1963–1969.

HOLMGREN PK, HOLMGREN, BARNETT LC (1990): Index Herbariorum. Part I. Herbaria of the World. 8th ed. Regnum Vegetabile 120. New York Botanical Garden, New York (http://www.nybg.org/bsci/ih/ih.html)

HOSAKA K, BATES ST, BEEVER RE, CASTELLANO MA, COLGAN W III, DOMÍNGUEZ LS, NOUHRA ER, GEML J, GIACHINI AJ, KENNEY SR, SIMPSON NB, SPATAFORA JW, TRAPPE JM (2006). Molecular phyloge-netics of gomphoid-phalloid fungi with an establishment of a new subclass Phallomycetidae and two new orders. *Mycologia* 98:949–959.

IOSIFIDOU P, AGERER R (2002). Die Rhizomorphen von *Gastrosporium simplex* und einige Gedanken zur systematischen Stellung der Gastrosporiaceae (Hymenomycetes, Basidiomycota). *Feddes Rep* 113:11–23.

IOSIFIDOU P, RAIDL S (2006).*Clavariadelphus pistillaris* (L.) Donk + *Fagus sylvatica* L. Descr Ectomyc 9/10:21–25.

JÜLICH W (1984). Die Nichtblätterpilze, Gallertpilze und Bauchpilze. Kleine Kryptogamenflora II b/1. Stuttgart New York.

KÕLJALG U, LARSSON K-H, ABARENKOV K, NILSSON RH, ALEXANDER IJ, EBERHARDT U, ERLAND S, HØILAND K, KJØLLER R, LARSSON E, PENNANEN T, SEN R, TAYLOR AFS, TEDERSOO L, VRALSTAD T, URSING BM (2005). UNITE: a database providing web-based methods for the molecular identification of ectomycorrhizal fungi. *New Phytol* 166:1063̃1068

LARSSON K-H, LARSSON E, KÕLJALG U (2004). High phylogenetic diversity among corticioid homobasi-diomycetes. *Mycol Res* 108:983–1002.

MLECZKO P (2004a). *Cantharellus cibarius* Fr. + *Pinus sylvestris* L. Descr Ectomyc 7/8:11–20.

MLECZKO P (2004b). *Craterellus tubaeformis* (Bull.) Quél. + *Pinus sylvestris* L. Descr Ectomyc 7/8:29–36.

MONCALVO J-M, NILSSON RH, KOSTER B, DUNHAM SM, BERNAUER T, MATHENY PB, PORTER TM, MARGA-RITESCU S, WEIß M, GARNICA S, DANELL E, LANGER G, LANGER E, LARSSON E, LARSSON H-H, VILGALYS R (2006). The cantharelloid clade: dealing with incongruent gene trees and phylogenetic recon-struction methods. *Mycologia* 98:937–948.

MOORE LM, JANSEN AE, VAN GRIENSVEN LJLD (1989). Pure culture synthesis of ectomycorrhizas with *Cantharellus cibarius*. *Acta Bot Neerl* 38:273–278.

NILSSON RH, LARSSON K-H, LARSSON E, KÕLJALG U (2006). Fruiting body-guided molecular identifica-tion of root-tip mantle mycelia provides strong indications of ectomycorrhizal associations in two species of *Sistotrema* (Basidiomycota). *Mycol Res* 110:1426–1432.

OGAWA M (1984). Ecological characters of ectomycorrhizal fungi and their mycorrhizae. An intro-duction to the ecology of higher fungi. *Japan A R Quaterly* 18:305–314.

PALFNER G (2001). Taxonomische Studien an Ektomykorrhizen aus den *Nothofagus*-Wäldern Mittel-südchiles. Bibliotheca Mycologica Vol. 190, Cramer, Berlin Stuttgart

PALFNER G, HORAK E (2001). *Gautieria inapire* sp. nov., a new hypogeous species from *Nothofagus* forest in Central Chile. *Sydowia* 53:140–151.

RAIDL S, AGERER R (1998). *Hysterangium stoloniferum* Tul. & C. Tul. + *Picea abies* (L.) Karst. *Descr Ectomyc* 3:31–35.

SCATTOLIN L, RAIDL S (2006). *Ramaria flavo-saponarea* R. H. Petersen + *Fagus sylvatica* L. *Descr Ectomyc* 9/10:135–141.

TEDERSOO L, KÕLJALG U, HALLENBERG N, LARSSON K-H (2003). Fine scale distribution of ectomycorrhi-zal fungi and roots across substrate layers including coarse woody debris in a mixed forest. *New Phytol* 159:153–165.

TEDERSOO L, SUVI T, E. LARSSON E, KÕLJALG U (2006). Diversity and community structure of ecto-mycorrhizal fungi in a wooded meadow. *Mycol Res* 110:734–748.

TRAPPE M, EBERHART J, LUOMA D (2000). *Craterellus tubaeformis* (Fr.) Quél. + *Tsuga heterophylla* (Raf.) Sarg. In: Goodman DM, Durall DM, Trofymow JA, Berch SM (eds) Concise Descriptions of North American Ectomycorrhizae. Mycologue Publications, and Canada-B. C. Forest Resource Development Agreement, Canadian Forest Service, Victoria, B. C., pp CDE23.1–CDE23.4

VILGALYS R, HESTER M (1990). Rapid genetic identification and mapping of enzymatically amplified ribosomal DNA from several *Cryptococcus* species. *J Bacteriol* 172:4238–4246.

WHITE TJ, BRUNS TD, LEE S, TAYLOR J (1990). Amplification and direct sequencing of fungal ribo-somal RNA genes for phylogenetics. In: Innis MA, Gelfand DH, Sninsky JJ, White TJ (eds) PCR Protocols, a Guide to Methods and Applications. Academic Press, San Diego, USA, pp 315–322.

ZAK B (1973). Classification of ectomycorrhizae. In: Marks G, Kozlowski TT (eds) Ectomycorrhi-zae, their ecology and physiology. Ac Press, New York, London, pp 43–78.

Figure legends

Fig. 1a, b Habit of monopodial-pyramidal mycorrhizal systems with many rhizomorphs. *(Fig. from RA 14583a).* **Fig. 2** Outer mantle layer with irregularly inflated hyphae, containing oil droplets, embedded in a slightly gelatinous matrix. Hyphae mostly simple septate, one large clamp evident. *(Fig. from RA 14583a).***Fig. 3** Middle mantle layer with irregularly inflated, simple septate hyphae with oil droplets, em-bedded in a matrix. *(Fig. from RA 14583a).***Fig. 4** Inner mantle layer with irregularly inflated, simple septate hyphae with oil droplets, embed-ded in a matrix. *(Fig. from RA 14583a).* **Fig. 5** Mantle surface of very tip with hyphae embedded in a slightly gelatinous matrix, clamps and simple septa as well as oil droplets within hyphae. *(Fig. from RA 14583a).* **Fig. 6** Surface of a thicker rhizomorph with a slightly gelatinous matrix, simple septa as well as large clamps, and with some ampullate inflations. *(Fig. from RA 14583a).* **Fig. 7** Optical section (below) through a thicker rhizomorph with the typical ampullate inflations at the septa or under incorporation of clamps; hyphae embedded in a gelatinous matrix; surface of the rhizomorphs (above). *(Fig. from RA 14583a).* **Fig. 8** Surface of a thin rhizomorph and ampullately inflated portions of emanating hyphae. *(Fig. from RA 14583a).***Fig. 9** Resupinate fruitbody with reticulate, shallowly poroid hymenophore. *Scale bar*=5 mm. *(Fig. from RA 14583).* **Fig. 10** Resupinate fruitbody with reticulate, shallowly poroid hymenophore. *Scale bar*=1 mm. *(Fig. from RA 14583)*

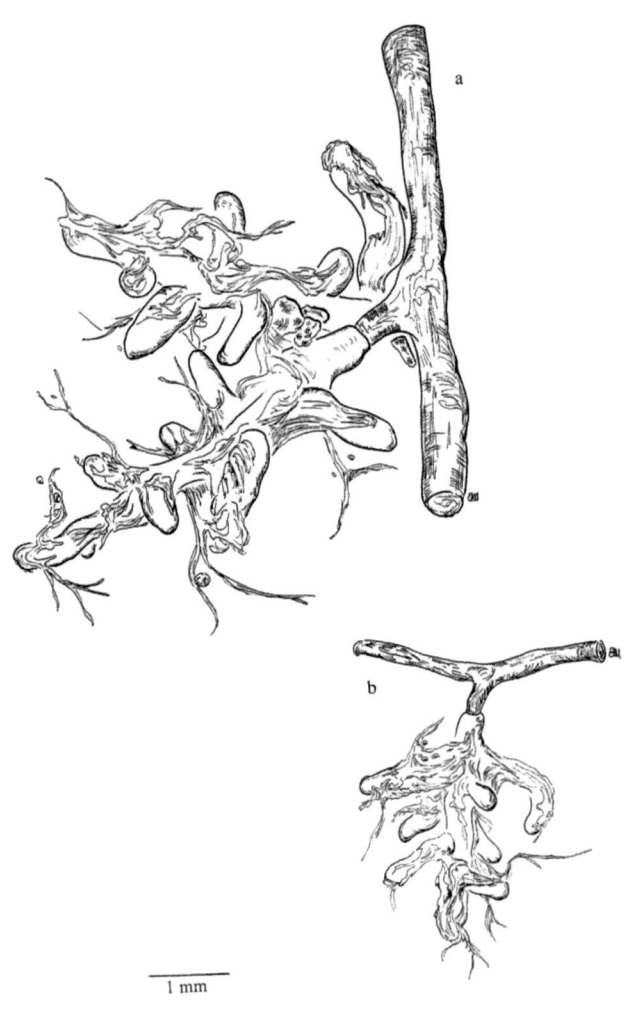

1 mm

Figs. 1a, 1b.

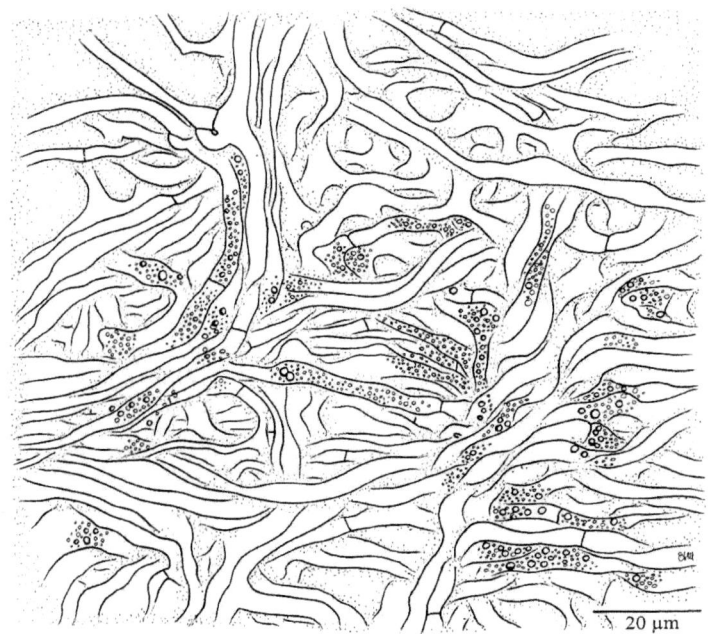

Fig. 2.

Fig. 3.

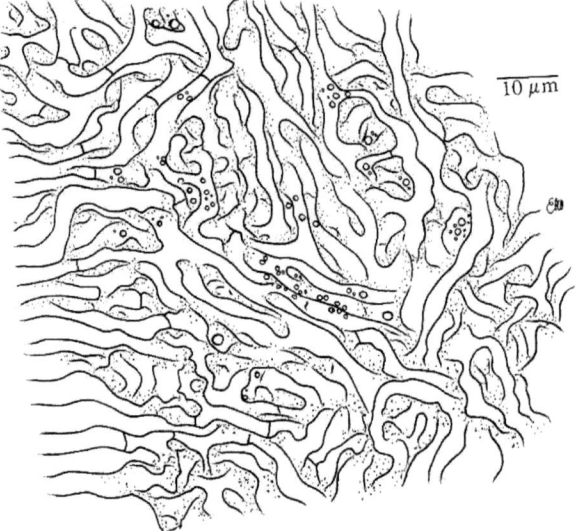

Fig. 4.

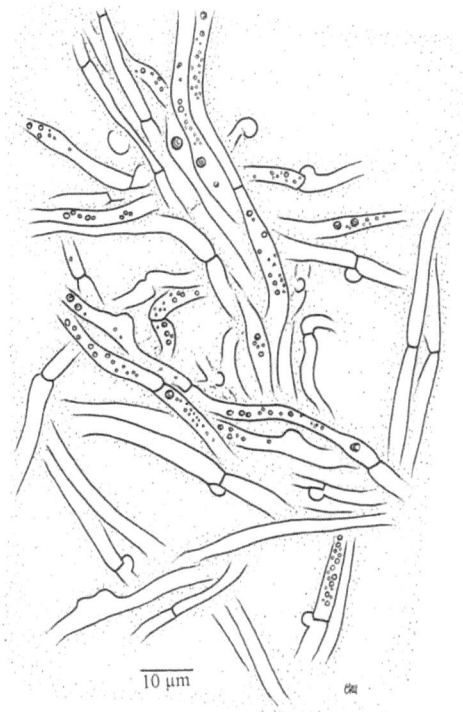

Fig. 5.

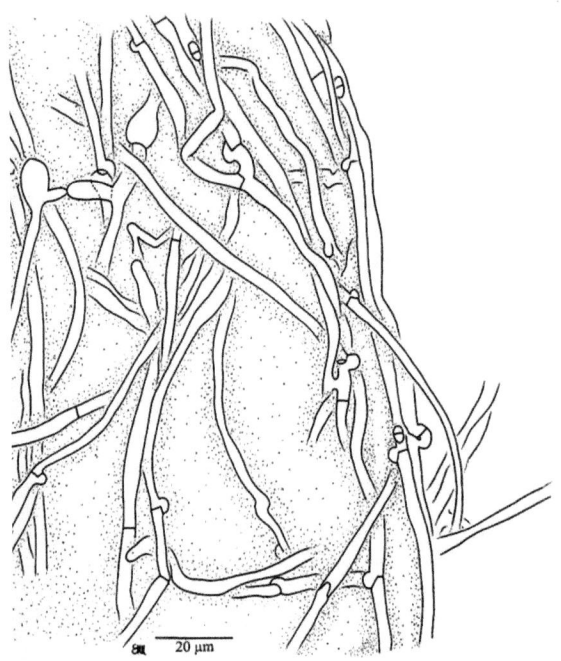

Fig. 6.

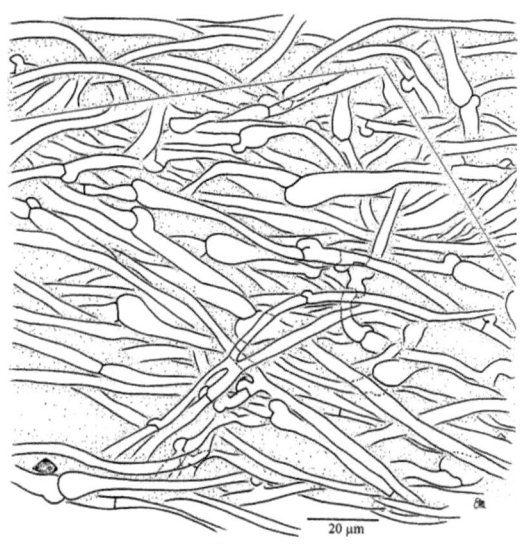

Fig. 7.

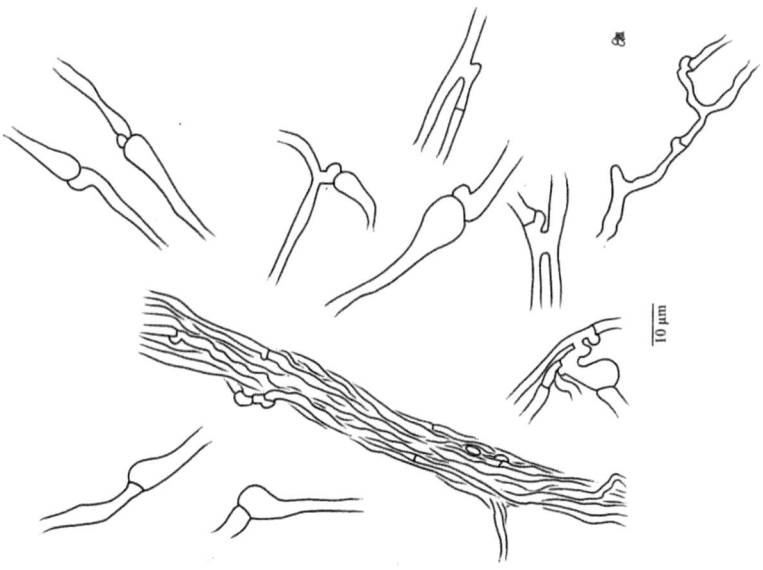

Fig. 8.

CHAPTER 9

The ectomycorrhizae of *Pseudotomentella humicola* on *Picea abies*

Erika Di Marino[1,3], Urmas Kõljalg[2] & Reinhard Agerer[3]

[1] *Dipartimento Territorio e Sistemi Agro-Forestali, Università degli Studi di Padova, viale dell'Università 16, I-35020 Legnaro, Padova, Italy*
[2] *Institute of Botany and Ecology, University of Tartu, 40 Lai Str., 51005 Tartu, Estonia*
[3] *Department Biology I, Biodiversity Research: Mycology, University of München, Menzinger Str. 67, D-80638 München, Germany*

Di Marino, E., U. Kõljalg & R. Agerer (2006) The ectomycorrhizae of *Pseudotomentella humicola* on *Picea abies*. – Nova Hedwigia 84: 429-440.

Abstract: Thelephoraceae are important contributors to the ectomycorrhizal diversity of temperate and boreal forests. Apart from *Peudotomentella tristis* and *P. larsenii*, *P. humicola* is now the third species of this genus, proven to form ectomycorrhizae. It forms brownish symbiotic organs when young that become dark brown to black at maturity; a slightly greenish colour appears when treated with KOH. The plectenchymatous, mantle forms ring- to star-like surface patterns with a gelatinous matrix between the hyphae and lacks cystidia. Rhizomorphs are dark brown to black, characterized by nodia and conical side-branches and are tightly enveloped by a layer of thin, irregularly and repeatedly branched hyphae. The differences to the ectomycorrhizae of *P. tristis* and to those of *Tomentella* spp. with plectenchymatous mantles and lacking cystidia are discussed.

Key Words: ectomycorrhizae, anatomy, DNA sequences, tomentelloid, thelephoroid

1. Introduction

Ectomycorrhizae (ECM) are considered as crucial for tree growth due to their powerful uptake of nutrients and water and their transport to tree roots of e.g. temperate and boreal forests (Read &

Smith, 1997). Many features of the ECM are functionally important. Mantles of the ECM may provide a shelter against microbial attack (Werner *et al.*, 2002) and might be a buffer against rapid loss of water, when the mantle hyphae form a gelatinous matrix (Agerer 2006), or they may provide a suitable surface for bacteria (Mogge *et al.*, 2000; Schelkle *et al.*, 1996; Timonen *et al.*, 1998) that might be helpful for the formation of ECM (Garbaye & Dopunnois, 1993) or for fixation of nitrogen (Amaranthus *et al.*, 1990). Smooth and hydrophilic mantles can directly acquire water and nutrients, while hydrophobic ECM with well developed rhizomorphs can transport nutrients over distances of several decimetres (Agerer 2001; Kammerbauer *et al.*, 1989; Schramm 1966; Skinner & Bowen, 1974). A great diversity of differentiation patterns can be discerned (Agerer 1987-2002; Agerer 1995; Agerer & Rambold, 2004-2005) and many of the characters are species-, or genus-specific (Agerer 2006). Others can characterize relationships of even higher systematic levels, i.e. order or superoders (Agerer 1999; 2006; Agerer & Iosifidou, 2004). The analysis and characterization of these functionally and ecologically crucial features is the goal of the functional anatomics of ECM. To unravel the potentials of ECM, these symbiotic organs of the more important species of different fungal relationships have to be studied in detail. In recent years, it have become more and more evident that ECM are frequently formed by the Basidiomycote order Thelephorales (Jakucs *et al.*, 2005; Kõljalg *et al.*, 2000; 2001; 2002). Here we describe the ECM of *Pseudotomentella humicola* M. J. Larsen, a member of the family Thelephoraceae.

2. Materials and Methods

The methods for characterization of ECM applied here are comprehensively described by Agerer (1991c). Fresh material was studied regarding morphology, colour of hyphae, and chemical reactions; material fixed in FEA (Agerer 1991) was applied to produce mantle and rhizomorph preparations as well as for longitudinal sections. The drawings were made using a ZEISS Axioskop with Normarski's Interference Contrast, at a magnification of 1300x with the aid of a drawing mirror, transferred on a transparent paper by Indian ink drawing devices, and finally reduced to a magnification of 1000x. Identification was possible by the association of the ECM with the fruitbodies and by tracing hyphal connections. In addition a comparison of nuclear rDNA ITS sequences obtained from the ECM root tip and from the fruitbody of *Pseudotomentella humicola* was performed. Methods of the DNA extraction, PCR and sequencing follows Tedersoo et al. (2006). EMBL accession numbers of the sequences are AM490945 and AM490946. The reference specimens of the ECM are deposited in M, the fruitbody in TU (Holmgren *et al.*, 1990). Collection data of the material: Norway, Akershus, Nannestad, 2 Sept. 2004, leg et det U. Kõljalg 100002 (in

TU).

3. Results

Morphological characters (Figs. 1a, b): *Mycorrhizal systems* abundant, dense, monopodiale-pyramidal, with distinct mantle surface; cortical cells not visible and mantle not transparent; mycorrhizal surface shiny, not smooth. - *Main axes* 7-8 mm long and 0.4 mm diam., straight. - *Unramified ends* up to 5 mm long and 0.3-0.4 mm diam., not inflated, cylindric. - *Surface of unramified ends* brown to black; older parts black, very tip ochre, yellowish brown or yellowish, not smooth, densely grainy or warty. – *Rhizomorphs* infrequent 20-60 µm diam., round in cross-section, originating proximally; concolorous to the mantle, brown or dark brown, connection to the mantle distinct; margin of rhizomorphs smooth; frequently ramified at restricted points; dimorphism not observed. – *Sclerotia* not observed.

Anatomical characters of mantle in plan views (Fig. 2): *Lacking are* cells densely filled with oily droplets or cells homogeneously filled with brownish contents, blue granules, needle-like contents, clamps, drops of exuded pigment. - *Outer mantle layers* (Figs. 2, 3, 6b) densely plectenchymatous, with ring-like arranged hyphal bundles (often rather star-like; mantle type A, Agerer 1991; 1995; Agerer 1987-2002, Agerer & Rambold, 2004-2005) and a gelatinous matrix; hyphae cylindric, not constricted at septa, septa very infrequent and hardly to discern; angles between hyphal junctions ca. 45° and less, membranaceously yellowish to brownish, smooth, cells walls up to 0.5 µm thick; cells 2-3(6) µm diam., 8-33 µm long; star-like structures of mantle surface with peripheral acuminate hyphal cells, cells of star centre glued by a dense matrix, hyphae of the stars (1.5)2-4(5) µm diam., 15-30(45) µm long. - *Middle mantle layers* (Fig. 4a) densely plectenchymatous, without pattern, membranaceously yellowish to brownish, cell walls up to 0.5 µm, smooth. - *Inner mantle layers* (Fig. 4b) densely plectenchymatous without pattern, membranaceously yellowish to brownish, cells 1.5-4(5.5) µm., cells (5)10-30(48) µm long.

Anatomical characters of emanating elements (Figs. 5b, 7b): Lacking are a gelatinous matrix, gelatinized hyphae, ampullate hyphae, drops of secreted pigment, and in IC strongly light reflecting crystals. - *Rhizomorphs* (Fig. 6a) 20-60 µm diam., rhizomorphs with nodia and conical structures at points of ramification, slightly differentiated (type C, Agerer 1991, 1995; Agerer 1987-2002, Agerer & Rambold 2004-2005; thelephoroid, Agerer 1999; Agerer & Iosifidou, 2004); surface covered by irregularly shaped, repeatedly ramified, densely entwining thin, apparently non-septate peripheral hyphae, hyphae 1-1.5 µm wide, membranaceously brownish to yellowish, smooth, thinner rhizomorphs (up to 30 µm) such hyphae lacking or covered only patchily; internal hyphae 2-3.5(4) µm diam., walls up to 0.5 µm wide, membranaceously brownish to yellowish, with clamps. -

Emanating hyphae not observed. – *Cystidia* not observed. - *Chlamydospores* not observed.

Anatomical characters, longitudinal section: Mantle 25–30(35) µm wide, plectenchymatous throughout, indistinctly 2-layered, outer mantle layer, comprising 3/4 - 4/5 of the mantle, with slightly elongate cells, inner mantle layer comprising 1/4 - 1/5 of the mantle, with roundish to somewhat cylindric cells; mantle at very tip (20)25-30(35) µm thick. - *Tannin cells* in 1(2) rows, asymmetrically-oval to -cylindric, in parallel orientation or slightly oblique to mycorrhizal surface, tangentially 65-70 µm, radially 5-7(10) µm; hyphal cells in one row, slightly beaded, cells 3-4(5) µm wide. - *Cortical cells* with Hartig net in 3(4) rows, asymmetrically-oval to -cylindric, oriented slightly oblique to mycorrhizal surface, reaching endodermis, tangentially (20)50-65(80) µm, radially (13)20-30(35) µm, CC_t = ca. 58 µm, CC_q = (1.2)2.4(4.2), hyphal cells in one row, cells 2-3 µm wide. - *Hartig net* in plan view of a sparsely lobed palmetti-type, lobes 2-2.5(3) µm wide, occasionally with septa.

Colour reaction with different reagents: Mantle and rhizomorph preparations: KOH 15%: outer mantle surface greenish; lactic acid: n.r. (no reaction); Melzer's reagent: possibly no reaction, but mantle too dark to test this feature .

Reference specimen for Picea ectomycorrhiza: Norway, Akershus, Nannestad, 2 Sept. 2004, leg. U. Kõljalg 100002, myc. isol. R. Agerer (in M).

Molecular analyses: fungal rDNA ITS sequences of the ECM root tip and *Pseudotomentella humicola* fruitbody were 100% identical.

4. Discussion

Pseudotomentella humicola is the second species of the genus where the ECM are described in detail. The first species was *P. tristis* (P. Karst.) M. J. Larsen (Agerer 1994). Lately a second species of the genus, *P. larsenii* Kõljalg & Dunstan (Kõljalg & Dunstan, 2001) was shown to form ECM. Further we will compare ECM formed by *P. tristis* and *P. humicola* because *P. larsenii* lacks ECM description in detail. Its ECM status with *Eucalyptus* sp. was confirmed by molecular analyses. Also, *P. larsenii* is found only in Australia while two other *Pseudotomentella* species are found in Northern hemisphere. Both, viz. *P. tristis* and *P. humicola* ECM, can be distinguished particularly by their colour. Whereas the ECM of *P. tristis* are blue and were growing on *Salix* those of *P. humicola* are dark brown to blackish and are formed on *Picea abies* (L.) Karst. Apart form these morphological differences the anatomy of both ECM differs considerably. *Pseudotomentella tristis* did not form rhizomorphs, in contrast to *P. humicola* that produces a slightly differentiated type with nodia and conical side-branches and it is, at least in older ontogenetical stages, densely covered by irregularly shaped, repeatedly ramified, densely entwining thin, apparently non-septate peripheral hyphae. Whereas the surface hyphae of the rhizomorphs are clampless, as are generally the mantle hyphae of *P. humicola* ECM, the internal hyphae bear clamps. Clamps are generally lacking in ECM of *P. tristis*. This is in agreement with the anatomy of fruitbodies: *Pseudotomentella humicola* forms clamps, they are completely lacking in *P. tristis* fruitbodies (Agerer 1994; Kõljalg 1996). The outer mantle layers are without any pattern in *P. tristis* (Agerer 1994), but furnished with distinct rings or even star-like structures in *P. humicola*. Both species have a gelatinous matrix between the mantle hyphae. Due to the dark mantle, tests for amyloidy are impossible in *P. humicola*, a feature characteristic of *P. tristis* ECM (Agerer 1994).

Both hitherto characterized *Pseudotomentella* species, belong to a group of Thelephoraceae that possesses plectenchymatous mantles and lacks cystidia on the mantle surface. All other species of this family form, as far as studied, either pseudoparenchymatous mantles and/or cystidia. Within the genus *Tomentella* plectenchymatous mantles and lacking cystidia apply for *T. ferruginea* (Pers.) Pat. (Raidl & Müller, 1996) and *Tomentella brunneorufa* M.J. Larsen (Agerer & Bougher, 2001). *Tomentellopsis submollis* (Svrcek) Hjortstam, the only species of this genus where ECM have been characterized to date, fits also to this combination of characters. All five mentioned species can well be kept apart with respect to their mantle structure and the organization of their extramatrical mycelium. Exclusively *P. tristis* does not form rhizomorphs on ECM, although the fruitbodies can sometimes possess them, but these rhizomorphs are delineated as dimitic, composed of thicker hyphae and of long, non-septate and apparently unbranched thin ones (Kõljalg 1996). *Tomentella brunneorufa* is characterized by the same type of rhizomorph as in *P. tristis* fruitbodies, whereas *P. humicola* and *T. ferruginea* are furnished by rhizomorphs that reveal nodia, conical side-branches and repeatedly ramified, densely entwining thin, apparently non-septate peripheral hyphae. *Tomentellopsis submollis*, however, forms only undifferentiated, uniform-compact rhizomorphs (Agerer 1999; Agerer & Iosifidou 2004; Agerer 2006).

Pseudotomentella humicola is distinct from *Tomentella ferruginea* by the ring- to star-like arrangement of outer mantle hyphae. This contrasts to the plectenchymatous mantle without any pattern of *T. ferruginea*. In addition, only *P. humicola* forms a gelatinous matrix between the ring-like arranged mantle hyphae. The leading hyphae (Agerer 1999) of *T. ferruginea* rhizomorphs bear bar-shaped crystals and the hyphae are covered by fine warts (Raidl & Müller, 1996). Both features are lacking in *P. humicola*.

Ring- or star-like arranged hyphae on the mantle surface are also characteristic of ECM of Bankeraceae (Agerer 2006). Star-like patterns were found on ECM of *Bankera fuligineoalba* (J.C. Schmidt) Coker & Beers ex Pouzar (Agerer & Otto 1997; 1998) and *Phellodon niger* (Fr.) P. Karst. (Agerer 1992a, 1993a), whereas ring-like or indistinctly ring-like patterns occur on mantles of

Boletopsis leucomelaena (Pers.) Fayod (Agerer 1992b; 1993b), *Hydnellum caeruleum* (Hornem.) P. Karst. (Kernaghan 2001); *H. peckii* Banker (Agerer 1993c, d), and *Sarcodon imbricatus* (L.) P. Karst. (Agerer 1991a, b). The rhizomorphs of these species differ considerably in structure from those of *Pseudotomentella humicola*. *Bankera fuligineoalba*, *Hydnellum caeruleum*, *H. peckii*, and *Phellodon niger* reveal uniform-compact rhizomorphs without any differentiation, clamps are lacking. For *Boletopsis leucomelaena* and *Sarcodon imbricatus*, however, phlegmacioid rhizomorphs (Agerer 1999, Agerer & Iosifidou 2004) are reported. Both species possess clamps. An important difference to *P. humicola* ECM is the carbonisation (Agerer 1987-2002; Agerer & Rambold, 2004-2005) of the ECM of *H. peckii*, *Phellodon niger*, and *Bankera fuligineo-alba*.

ECM of Bankeraceae are known to produce chlamydospores (Agerer 2006; Agerer & Rambold, 2004-2005), a feature lacking in *P. humicola*, but occurring in a few *Pseudotomentella* species, in *P. atrofusca* M. J. Larsen (Kõljalg 1996; Melo *et al.*, 2002), *P. rhizopunctata* Martini & Hentic (Martini & Hentic, 2003), and P. *vepallidospora* M. J. Larsen (Kõljalg 1996). But chlamydospores are, with exception of *Tomentella guadalupensis* Martini & Hentic (Martini & Hentic, 2005), unknown in *Tomentella* fruitbodies and ECM (Larsen *et al.*, 1994; Melo *et al.*, 2000; 2002; 2003; 2006; Kõljalg 1996). The identical very specialized structure of the chlamydospores of *T. guadalupensis* (Martini & Hentic, 2005) and those of *Hydnellum peckii* (Agerer 1993c) perhaps indicates a closer relationship between these two species, as for Agaricales, Walther *et al.* (2005) could show that anamorphs provide valuable characters for a natural classification. But considerably more studies have to focus on ontogeny and final structure of thelephoralean chlamydospores.

Two further, but still unidentified ECM are suggested to belong to Thelephoraceae or Bankeraceae and indicate certain similarities to *Pseudotomentella humicola*. *Pinirhiza discolor* (Golldack *et al.*, 1998) mantles are undifferentiated, hyphae are clampless, and form chlamydospores on slightly differentiated rhizomorphs, *Pinirhiza stellannulata* (Golldack *et al.*, 1996) forms ring- to star-like mantles, undifferentiated rhizomorphs, and lack clamps.

In conclusion, the ECM of *Pseudotomentella humicola* can be clearly distinguished from any other ECM comprehensively described to date by the presence of star-like hyphal mantle pattern, by a gelatinous matrix, by lacking cystidia, by nodia and conical side-branches forming rhizomorphs that are tightly enveloped by a layer of thin, irregularly and repeatedly branched hyphae. It requires further studies whether the rhizomorphs of *P. humicola* are convergently evolved to identically shaped ones of some *Tomentella* species (e. g. *T. ferruginea*) or whether this special rhizomorphal structure could indicate a closer relationship.

Acknowledgments

We are very much obliged to Norfa, who supported our stay in Norway (UK and RA) during a field course for PhD students, during that the ectomycorrhizae and fruitbody of *Pseudotomentella humicola* could be found and isolated.

References

AGERER R (1987-2002). Colour Atlas of ectomycorrhizae. 1st-12th delivery. Einhorn-Verlag, Schwäbisch Gmünd.

AGERER R (1991a). Ectomycorrhizae of *Sarcodon imbricatus* on Norway spruce and their chlamydospores. Mycorrhiza 1: 21-30.

AGERER R (1991b). *Sarcodon imbricatus*. In Agerer R (ed.) Colour Atlas of Ectomycorrhizae, plate 66. Einhorn, Schwäbisch Gmünd.

AGERER R (1991c). Characterization of ectomycorrhiza. In Norris JR, Read DA, Varma AK (eds) Techniques for the study of mycorrhiza. Methods in Microbiology, vol 23, Academic Press, London et al., pp 25-73.

AGERER R (1992a). Ectomycorrhizae of *Phellodon niger* on Norway spruce and their chlamydospores. Mycorrhiza 2: 47-52.

AGERER R (1992b). Studies on ectomycorrhizae XLIV. Ectomycorrhizae of *Boletopsis leucomelaena* (Thelephoraceae, Basidiomycetes) and their relationship to an unidentified ectomycorrhiza. Nova Hedwigia 55: 501-518.

AGERER R (1993a). *Phellodon niger*, In: Agerer R (ed) Colour Atlas of Ectomycorrhiza, plate 78, Einhorn, Schwäbisch Gmünd.

AGERER R. (1993b). *Boletopsis leucomelaena*. In: Agerer R (ed) Colour Atlas of Ectomycorrhizae, plate 75. Einhorn, Schwäbisch Gmünd.

AGERER R (1993c). Ectomycorrhizae of *Hydnellum peckii* on Norway spruce and their chlamydospores. Mycologia 85: 74-83.

AGERER R (1993d). *Hydnellum peckii*. In: Agerer R (ed) Colour Atlas of Ectomycorrhizae, plate 77. Einhorn, Schwäbisch Gmünd.

AGERER R (1994). *Pseudotomentella tristis* (Thelephoraceae). Eine Analyse von Fruchtkörper und Ektomykorrhizen. Z Mykol 60: 143-158.

AGERER R (1995). Anatomical characteristics of identified ectomycorrhizas: an attempt towards a natural classification. In Varma K, Hock B (eds) mycorrhiza: structure, function, molecular biology and biotechnology. Springer, Berlin, Heidelberg, New York, pp 685-734.

AGERER, R (1999). Never change a functionally successful principle: the evolution of Boletales s. l. (Hymenomycetes, Basidiomycota) as seen from below-ground features. *Sendtnera* 6: 5-91.

AGERER R (2001). Exploration types of ectomycorrhizae. A proposal to classify ectomycorrhizal mycelial systems according to their patterns of differentiation and putative ecological importance. *Mycorrhiza* 11(2): 107-114.

AGERER R (2006). Fungal relationships and structural identity of their ectomycorrhizae. *Mycol Progress* 5 (in print).

AGERER R, BOUGHER NL (2001). *Tomentella brunneorufa* M. J. Larsen + *Eucalyptus* spec. *Descr Ectomyc* 5: 205-212.

AGERER R, IOSIFIDOU P (2004). Rhizomorph structure of Hymenomycetes: a possibility to test DNA-based phylogenetic hypotheses? In Agerer, R., Piepenbring, M. & P. Blanz (eds.) Frontiers in Basidiomycote Mycology, IHW-Verlag, Eching, pp 249-302.

AGERER R, OTTO P (1997). *Bankera fuligineo-alba* (J. C. Schmidt.: Fr.) Pouzar + *Pinus sylvestris* L. *Descr Ectomyc* 2: 1-6.

AGERER R, OTTO (1998) P. *Bankera fuligineo-alba*. In: Agerer R (ed) Colour Atlas of Ectomycorrhizae, plate 115. Einhorn, Schwäbisch Gmünd.

AGERER R, RAMBOLD G (2004 – 2005 [first posted on 2004-06-01; most recent update: 2005-10-25]). DEEMY – An information system for characterization and DEtermination of EctoMYcorrhizae. www.deemy.de – München, Germany.

AMARANTHUS MP, LI CY, PERRY DA(1990). Influence of vegetation type and madrone soil inoculum on associative nitrogen fixation in Douglas-fir rhizosphere. *Can J For Res* 20: 368-371.

GARBAYE J, DUPONNOIS R (1993). Specificity and function of mycorrhization helper bacteria (MHB) associated with *Pseudotsuga menziesii Laccaria laccata* symbiosis. *Symbiosis* 14(1-3): 335-344.

GOLLDACK J, MÜNZENBERGER B, AGERER R, HÜTTL RF (1996). "*Pinirhiza stellannulata*" + *Pinus sylvestris* L. *Descr Ectomyc* 1: 89-93.

GOLLDACK J, MÜNZENBERGER B, HÜTTL RF (1998). "*Pinirhiza discolor*" + *Pinus sylvestris* L. *Descr Ectomyc* 3: 55-60.

HOLMGREN PK, HOLMGREN NH, BARNETT LC (1990). Index Herbariorum. Part I. Herbaria of the World. 8th edn. Regnum Vegetabile 120. New York Botanical Garden, New York (http://www.nybg.org/bsci/ih/ih.html).

JAKUCS E, KOVACS GM, AGERER R, ROMSICS C, ERÖS-HONTI Z (2005). Morphological-anatomical characterization and molecular identification of *Tomentella stuposa* ectomycorrhizae and related anatomotypes. *Mycorrhiza* 15: 247-258.

KAMMERBAUER H, AGERER R, SANDERMANN H (1989). Studies on ectomycorrhiza XXII. Mycorrhizal rhizomorphs of *Thelephora terrestris* and *Pisolithus tinctorius* in association with Norway spruce (Picea abies): formation in vitro and translocation of phosphate. *Trees* 3: 78-84.

KERNAGHAN G (2001). Ectomycorrhizal fungi at tree line in the Canadian Rockies II. Identification of ectomycorrhizae by anatomy and PCR. *Mycorrhiza* 10: 217-229.

KÕLJALG U (1996). *Tomentella* (Basidiomycota) and related genera in temperate Eurasia. Fungiflora, Oslo.

KÕLJALG U, DAHLBERG A, TAYLOR AFS, LARSSON E, HALLENBERG N, STENLID J, LARSSON KH, FRANSSON PM, KARÉN O, JONSSON L (2000). Diversity and abundance of resupinate thelephoroid fungi as ectomycorrhizal symbionts in Swedish boreal forests. *Molec Ecology* 9: 1985-1996.

KÕLJALG U, DUNSTAN W (2001): *Pseudotomentella larsenii* sp. nov. (Thelephorales), a common ectomycorrhiza former in dry eucalypt woodland and forests of Western Australia. *Harvard Papers in Botany* 6(1): 123-130.

KÕLJALG U, JAKUCS E, BOKA K, AGERER R (2001). Three ectomycorrhiza with cystidia formed by different *Tomentella* species as revealed by rDNA ITS sequences and anatomical characteristics. *Folia Cryptog. Estonia Facs.* 38: 27-39.

KÕLJALG U, TAMMI H, TIMONEN S, AGERER R, SEN R (2002). ITA rDNA sequence-based phylogenetic analysis of *Tomentellopsis* species from boreal and temperate forests, and the identification of pink-type ectomycorrhizas. *Mycol Progress* 1: 81-92.

LARSEN MJ, BELTRAN-TEJERA E, RODRIGUEZ-ARMAS JL (1994). *Tomentella oligofibula* sp. nov. (Aphyllophorales, Thelephoraceae s. str.), from the Canary Islands. *Mycotaxon* 52: 109-112.

MARTINI C, HENTIC R (2003). *Pseudotomentella rhizopunctata* sp. nov., une nouvelle espèce de champignon tomentelloide chlamydosporée. *Bull Soc Mycol Fr* 119(1-2): 19-29.

MARTINI EC, HENTIC R (2005). *Tomentella lilacinogrisea* et *T. guadalupensis* sp. nov. Deux espèces de champignons tomentelloides des caraibes. *Bull Soc Mycol Fr.* 121: 17-27.

MELO I, SALCEDO I (2002). Contribution to the knowledge of tomentelloid fungi in the Iberian Peninsula. III. *Nova Hedwigia* 74: 387-404.

MELO I, SALCEDO I, TELLERÍA MT (2000). Contribution to the knowledge of tomentelloid fungi in the Iberian Peninsula. II. *Karstenia* 40: 93-101.

MELO I, SALCEDO I, TELLERÍA MT (2003). Contribution to the knowledge of tomentelloid fungi in the Iberian Peninsula. IV. *Nova Hedwigia* 77: 287-307.

MELO I, SALCEDO I, TELLERÍA MT (2006). Contribution to the knowledge of tomentelloid fungi in the Iberian Peninsula. V. *Nova Hedwigia* 82: 167-187.

MOGGE B, LOFERER C, AGERER R, HUTZLER P, HARTMANN A (2000). Bacterial community structure and colonization patterns of Fagus sylvatica L. ectomycorrhizospheres as determined by fluorescence in situ hybridization and confocal laser scanning microscopy. *Mycorrhiza* 9: 271-278.

RAIDL S, MÜLLER WR (1996). *Tomentella ferruginea* (Pers.) Pat. + *Fagus sylvatica* L. *Descr Ectomyc* 1: 161-166.

SCHELKLE M, URSIC M, FARQUHAR M, PETERSON RL (1996). The use of laser scanning confocal microscopy to characterize mycorrhizas of *Pinus strobus* L. and to localize associated bacteria. *Mycorrhiza* 6: 431-440.

SCHRAMM JR (1966). Plant colonization studies on black wastes from anthracite mining in Pennsylvania. *Trans Am Philos Soc* 56: 5-189.

SKINNER MF, BOWEN GD (1974). The uptake and translocation of phosphate by mycelial strands of pine mycorrhizae. *Siol Biol Biochem* 6: 53-56.

SMITH SE, READ DJ (1997). *Mycorrhizal symbiosis*. 2nd ed. Academic Press, San Diego, London et al.

TEDERSOO L, SUVI T, LARSSON E, KÕLJALG U (2006). Diversity and community structure of ectomycorrhizal fungi in a wooded meadow. *Mycol Res* 110: 734-748.

TIMONEN S, JØRGENSEN K, HAAHTELA K, SEN R (1998). Bacterial community structure at defined locations of the *Pinus sylvestris* - *Suillus bovinus* and *Pinus sylvestris* - *Paxillus involutus* mycorrhizospheres in dry pine forest humus and nursery peat. *Can J Microbiol* 44: 499-513.

WALTHER G, GARNICA S, WEIß M (2005). The systematic relevance of conidiogenesis modes in the gilled Agaricales. *Mycol Res* 109: 525-544.

WERNER A, ZADWORNY M, IDZIKOWSKA K (2002). Interaction between *Laccaria laccata* and *Trichoderma virens* in co-culture and in the rhizosphere of *Pinus sylvestris* grown in vitro. *Mycorrhiza* 12: 139-145.

Captions: **Fig. 1a, b.** Habit. - **Fig. 2.** Outer mantle surface: net with ring to star-like arranged hyphal bundles imbedded in a gelatinous matrix, septa very infrequent and hardly discernible. - **Fig. 3.** Outer mantle layer: densely plectenchymatous. - **Fig. 4a.** Middle mantle layer: densely plectenchymatous, without pattern. - **Fig. 4b.** Inner mantle layer: densely plectenchymatous without pattern. - **Fig. 5a, b.** Young rhizomorphs: surface partially covered by irregularly shaped, apparently non-septate, thin peripheral hyphae; with indistinct nodia. - **Fig. 6a.** Young, slightly differentiated rhizomorph, with clamped hyphae and a few thin peripheral hyphae. - **Fig. 6b.** Star of mantle surface: stars with peripheral acuminate hyphal cells, cells of star centre glued by a dense matrix. - **Fig. 7a.** Conical side branch of an older rhizomorph covered by irregularly shaped, repeatedly ramified, densely entwining thin, apparently non-septate peripheral hyphae. – **Fig. 7b.** Older rhizomorph with internal clamped hyphae and covered by irregularly shaped, repeatedly ramified, densely entwining thin, apparently non-septate peripheral hyphae in surface view (below) and optical section (above).

Fig. 1

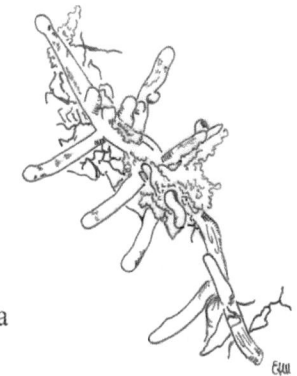

a

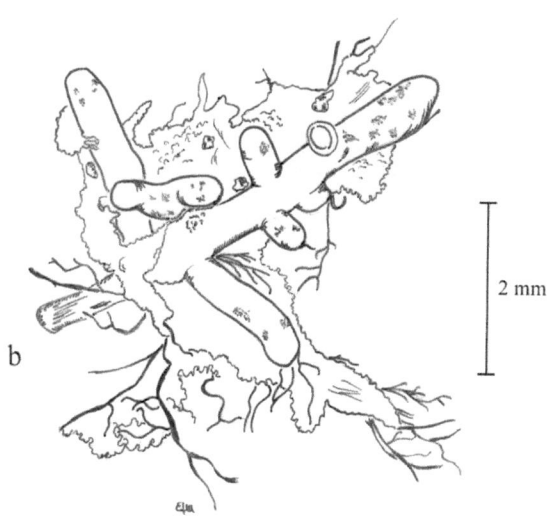

b

Fig. 2

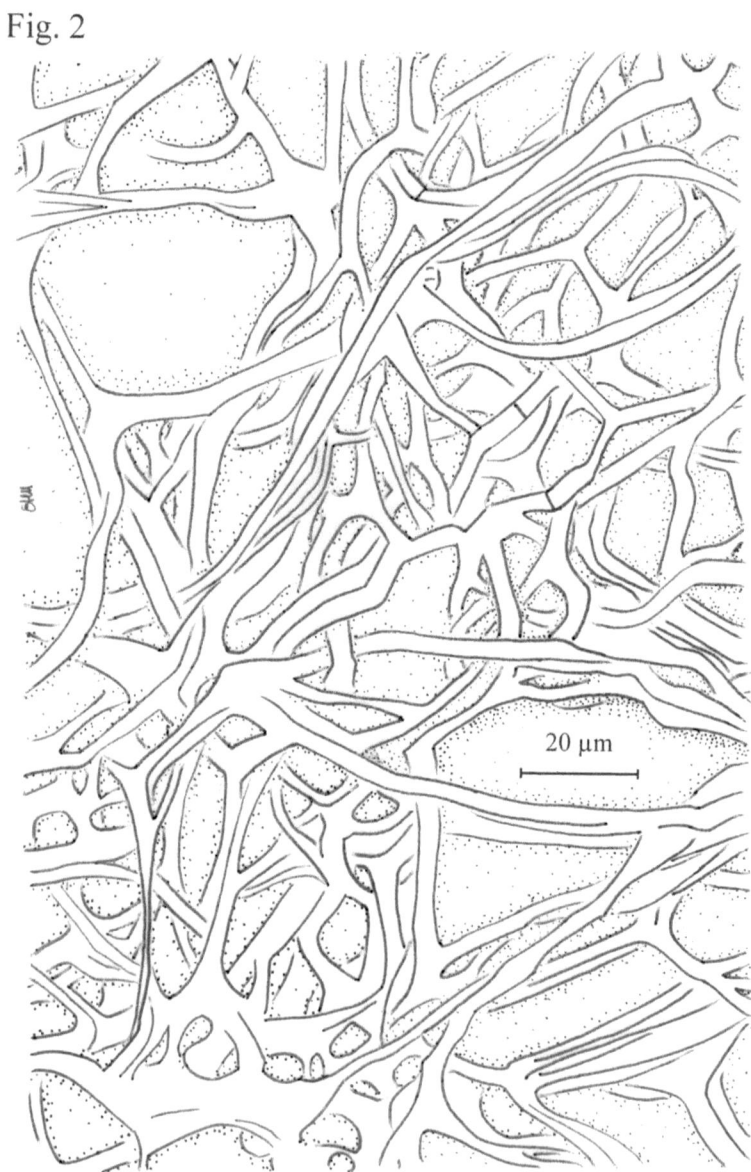

Fig. 3

Fig. 4

a

20 μm

b

Fig. 5

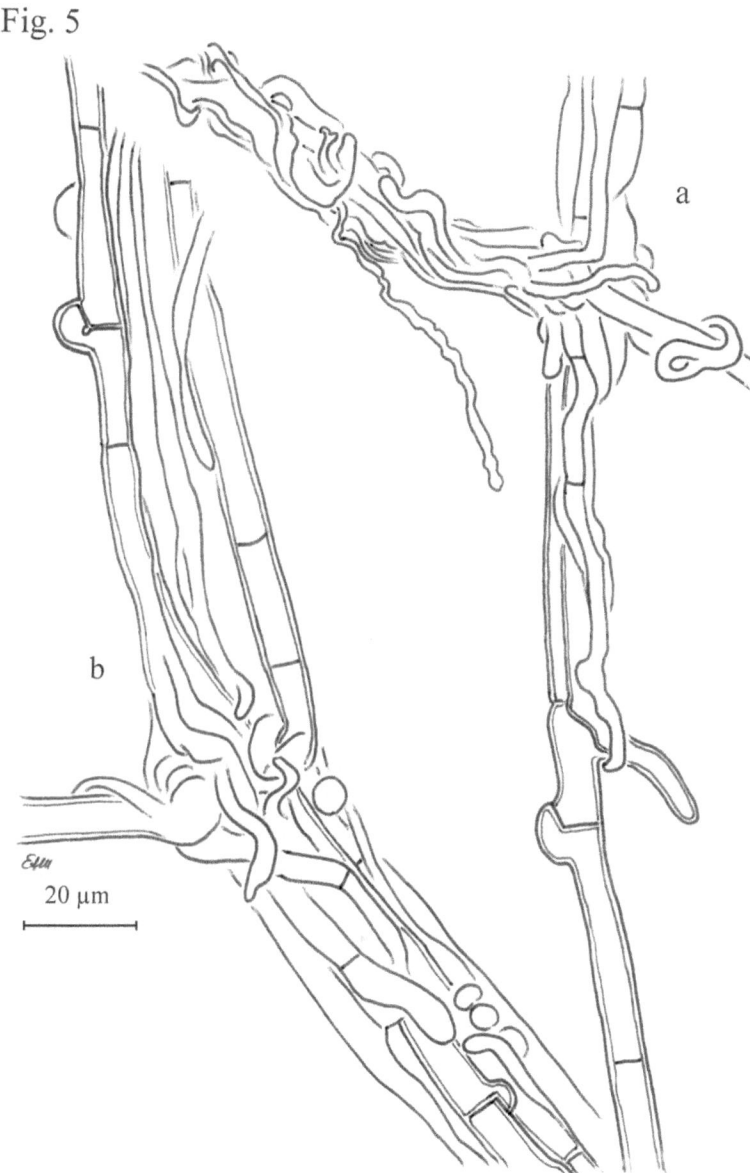

20 μm

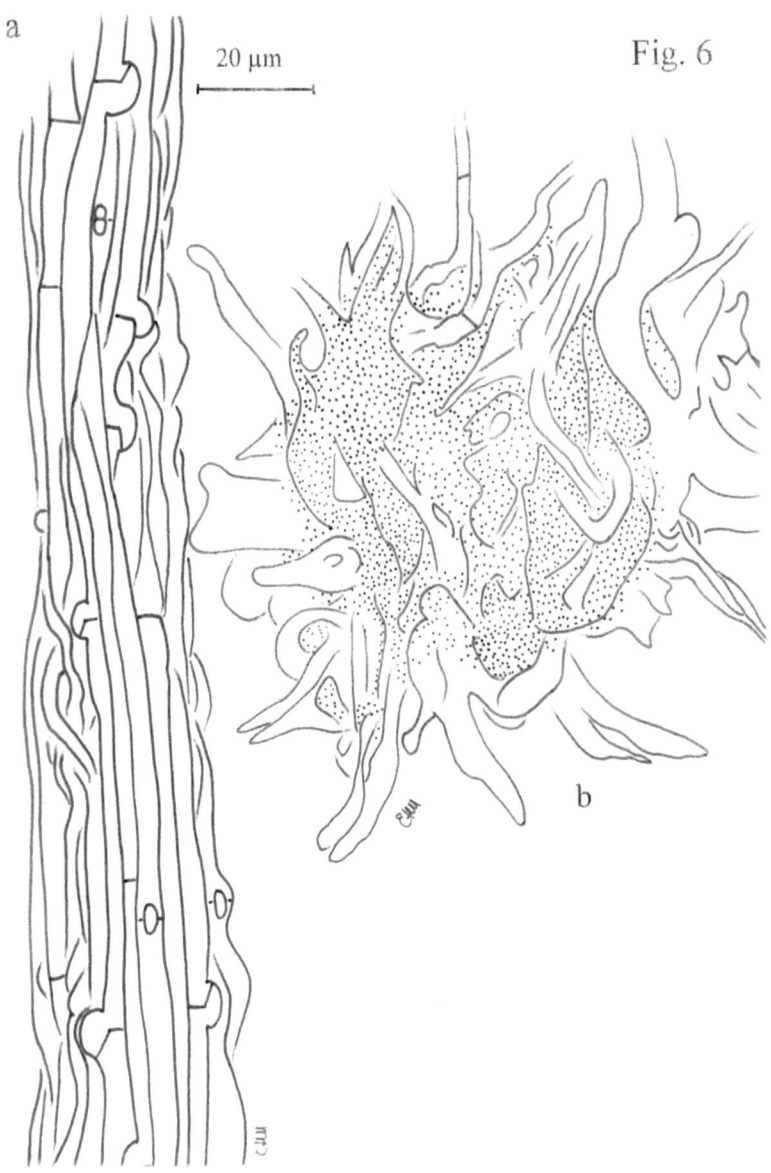

Fig. 6

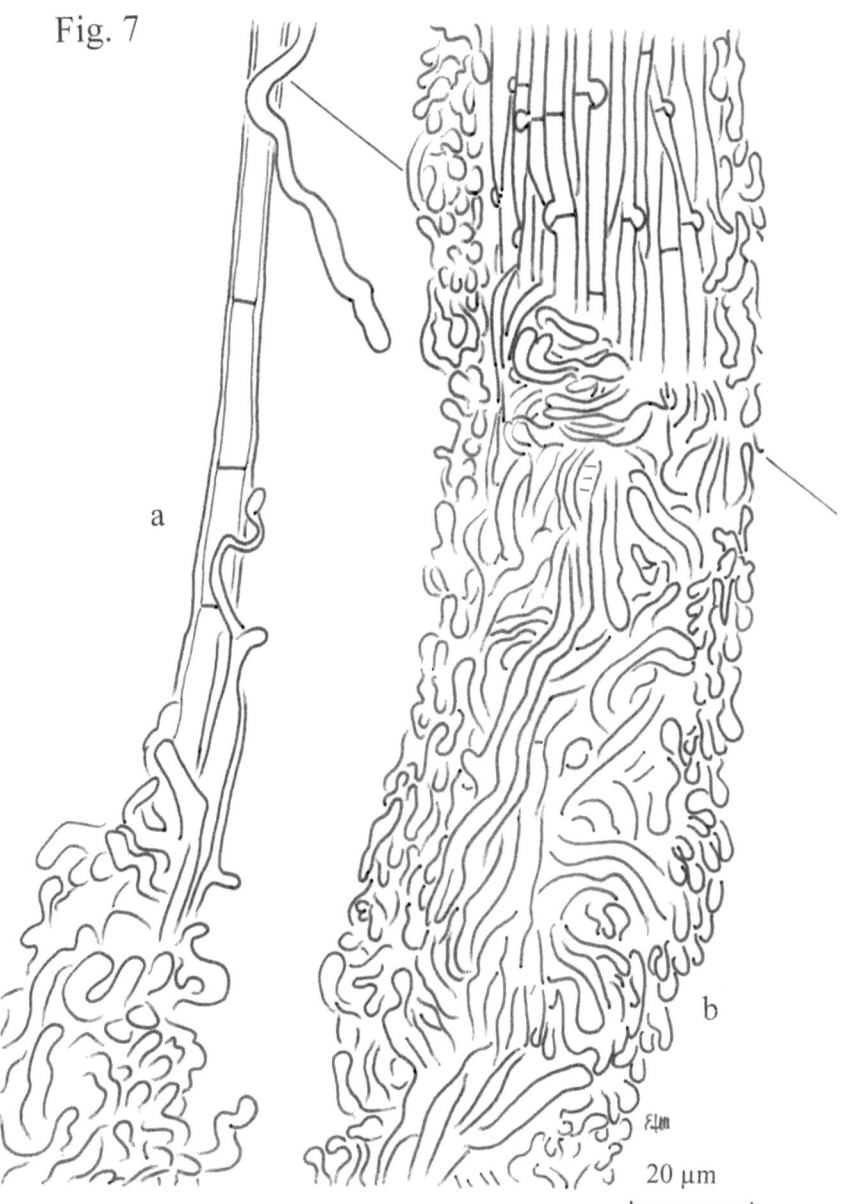

Fig. 7

20 μm

CHAPTER 10
Conclusion

The EM (ectomycorrhizae) communities can be strongly influenced by a range of forest management practices (reviewed by Jones et al., 2003) and in particular numerous field researches reported as harvesting outcomes declines of EM diversity. Harvesting significantly decreased the thickness of the humus layer as well as decreasing the number of EM root tips both metre root length and per unit humus volume, like to reported by Mahmood et al. (1999), in a swedish spruce forest.

Other sylvicultural practises have an impact on the main parameters of the EM community: i.e. the abundance and diversity of mycorrhizal fungi are generally negatively affected by clear-cutting, and its main disturbance effect on the community are quantified as a loss of inoculation potential n, or/and due to the decrease in inputs of carbon from host plants, or/and a combination of drastic changes in the environmental conditions (Orlander et al., 1990; Sutton 1993; Hagerman et al., 1999; Durall et al., 1999; Byrd et al., 2000; Cline et al., 2005). Also strong thinning can modify the ectomycorrhizal community structure as reported for old Beech stands (Buée et al., 2005) and in a declining pedunculate oak forest (Mosca et al., 2007).

The aim of these researches was to understand the possible effects of coppicing on the EM community structure in 7 different Beech stands but comparable (for the main stand features and the beech presence), chosen for their high productivity and for their very frequent utilization since the past. Following the objectives fixed by the Kyoto Protocol, the European Union and the Italian Government promoted actions towards the development of renewable resources (Bernetti et al., 2004). Short rotation coppices could be important instruments to enforce these policies. In these context the possibility to reduce the rotation also in Beech stands with sustainable effect on the ecosystem and without a loss of biodiversity could be a new opportunity of study.

In the present work the results confirmed the ectomycorrhizal community structure investigated in 7 beech coppices of different age was typical with respect of the occurrence of few abundant species and many others with significantly lower abundance as reported in precedent investigations (Grogan et al., 2000; Horton & Bruns, 2001; Taylor, 2002; Montecchio et al., 2004; Mosca et al., 2007; Scattolin et al., 2008). A dominance of Thelephoroid and Cortinareaceous fungi was also observed. This composition is well-known, because recent studies discussed the evidence of that EM are frequently formed by the Basidiomycote order Thelephorales (Jakucs et al., 2005; Kõljalg et al., 2000; 2001; 2002) and by the presence of *Cortinarius* species, (Kjøller 2006). *Cenococcum geophilum* was the most frequently detected species in each site and in each sample date, whether occurring in dolomitic and calcareous sites, probably due to its known ability to produce antifungal compounds active against several filamentous fungi (Koide et al., 2005) and to its high drought tolerance (Neves Machado, 1995; Jany et al., 2003). Morphological, anatomical and molecular investigations revealed a total of 60 anatomotypes. Of these 35 were unknown on *Fagus sylvatica* up to now (De Roman et al., 2005). 7 not described ectomycorrhizae were assigned to family or ordinal level *(Thelephorales, Boletales, Pezizales, Sebacinaceae, Thelephoraceae),* 19 to genus *(Amphinema* sp., *Boletus* sp., *Cortinarius* sp., *Craterellus* sp., *Entoloma* sp., *Hydnum* sp., *Hygrophorus* sp., *Inocybe* sp., *Laccaria* sp., *Lactarius* sp., *Ramaria* sp. *Sebacina sp.*, *Tomentella* sp.), 3 to species *(Cortinarius ionochlorus, Cortinarius infractus, Hygrophorus penarius),* 3 ectomycorrhizae not identified at present were described in detail *(Fagirhiza byssoporioides, Fagirhiza entolomoides, Fagirhiza stellata),* while 3 remained unidentified.

The achieved results demonstrated that along a wide coppice frequency gradient (2 to 48 years, with 25 years being the rule), the main EM community parameters, like tips' vitality and mycorrhization, changed only in the vertical distribution with a major abundance of EM not vital in the organic soil layers confirming only partially the work of other authors (Baier et al., 2006). Moreover the ecological indexes attested that the richness and evenness varied only on the temporal scale (related to the different collections), but they were not correlated with the coppice frequency or the slope, partly confirming available information from clear-cutting and thinning experiments (Buée et al., 2005; Cline et al., 2005; Mosca et al., 2007), and explainable with an hypothetical *resilience,* as an "adaptive diversity".

No relevant differences in the EM spatial and vertical distribution with the shoot age were revealed in the two years of the research, leading to the hypothesis that the coppice treatment in these Beech stands, did not have a significant and direct effect on the EM richness and community structure since 2 to 48 years from coppicing. In fact the multivariate analyses demonstrated that the EM presence, richness and distribution was never mainly associated with the shoot age, but strongly related to stands conditions like slope steepness and the soil moisture. Also the ecological features of the EM species like the hydrophobicity or the exploration types (Agerer 2001) didn't define particular conditions in the soil correlated to the coppices effect. Due the high presence of hydrophilic EM species equal distributed in all the sites, it was impossible to find a clear correlation between these ecological features and the coppicing frequency.

Furthermore, when the MID (Morisita index) was applied with significant results, the most frequent species always revealed an aggregated distribution (in the vertical distribution MID was always significant showing 39 species aggregated in the sites, for the first spatial collection in June 2005 it was significant for all species, in the second collection, in October 2005, this index was significant for only 23 species compared to the total of 46 species collected, while for the last collection in June 2006, it was significant for 19 species compared to 29 total species) supporting the hypothesis that micro-scale effects (i.e. antagonistic interactions among species) prevail on macro-scale features (i.e. humus and bedrock type, plant age) as previously shown (Bruns 1995; Toljander et al., 2006; Gebhardt et al., 2007).

Bruns (1995) in fact attested the very different physical-chemical situations present in a forest soil contribute to create this spatial heterogeneity, and it is involved in the maintenance of high ECM fungal diversity, as confirmed also by other works (Toljander et al., 2006; Gebhart et al., 2007). But only few studies have examined the micro-spatial distribution of individual ECM species in relation to soil factors, as a possible result of the ecosystem resilience (Toljander et al., 2006; Gebhart et al., 2007; Mosca et al., 2007; Scattolin et al., 2008).

The response of the EM community to the repeated manipulation of litter and humus layers, that was documented to be strongly through coppicing (Buckley 1992), needs more detailed investigations, because in this study no clear differences were found between the organic and mineral soil layers, contrasting previously results (Kuyper & Landeweert, 2002; Baier et al., 2006).

Taking into account the stability of the EM community as a possible indicator of plant health status (Wargo 1988; Fellner & Caisovà, 1994; Causin et al., 1996; Montecchio et al., 2004; Mosca et al., 2007, Scattolin et al., 2008), "Short rotation" practices in Beech forests could be considered a sustainable activity, according to the new trends in EU energetic policies, aimed to promote the increase of renewable energetic resources availability (Cutini 2001). From this point of view, new guidelines could be provided for the sylviculture management. For assessing ecosystem resilience within the context of the global change, the identification of the ecological features determining this "adaptive diversity" in EM communities, will have more and more importance (Dahlberg 2001).Further investigations to verify if and how a high and repeated coppice frequency can cause irreversible alterations in EM biodiversity are needed.

The results here reported, on the mycorrhizae of *Hygrophorus penarius* on *Fagus sylvatica* seems to be more important not only for the systematic aspects but also from the ecological point of view. This mycorrhizae was described and compared to other species, and showed a similar behaviour to that of *Entoloma saepium* on *Rosa* sp. Like *E. saepium*, *H. penarius* showed an attitude to digest the root meristem and the young root cells, in a parasitic-like activity. The Hartig net was not formed, although a very thick gelatinous mantle composed by infrequently clamped hyphae embedded in a very distinctive matrix providing the mycorrhiza with an almost transparent mantle was present. To get more information about its behaviour the stable carbon and nitrogen isotope ratios of its mycorrhizae were studied, revealing a negative $\delta^{15}N$ values, similar to that of non-mycorrhiza roots and of many typical ectomycorrhizae. $\delta^{13}C$ values did not reveal important information..

As previously suggested in other ectomycorrhizal symbioses (Schwacke & Hager 1992; Salzer *et al.*, 1996), the wide and unspecific host reactions induced by *H. penarius,* could be effective against other microorganisms, in accordance with well known "induced resistance" strategies (Sticher *et al.*, 1997; van Loon *et al.*, 1998).

Further anatomo-physiological analyses on *H. penarius* behaviour in different tip age and seasons, and on its potential ability to induced a non-specific plant resistance to possible parasites are therefore of main importance.

This work reports also 2 descriptions of new ectomycorrhizal species: *Pseudotomentella humicola* on *Picea abies* and *Sistotrema muscicola* on *Castanea sativa*. *P. humicola* is now the third species of this genus, proven to form ectomycorrhizae apart from *P. tristis* and *P. larsenii*, while the ectomycorrhizal status of *Sistotrema muscicola* is shown for the first time unequivocally, although already previously sistotremoid DNA had been extracted from ectomycorrhizae.

References

AGERER R (2001). Exploration types of ectomycorrhizae. A proposal to classify ectomycorrhizal mycelial systems according to their patterns of differentiation and putative ecological importance. *Mycorrhiza* 11: 107-114.

BAIER R, INGENHAAG J, BLASCHKE H, GÖTTLEIN A, AGERER R (2006). Vertical distribution of an ectomycorrhizal community in upper soil horizons of a young Norway spruce (*Picea abies* [L.] Karst.) stand of the Bavarian Limestone Alps. *Mycorrhiza* 16(6): 197-206.

BERNETTI I, FAGARAZZI C, FRATINI R (2004). A methodology to analyse the potential development of biomass-energy sector: an application in Tuscany. *Forest Policy and Economics* 6: 415-432.

BRUNS TD (1995). Thoughts on the processes that maintain local species diversity of ectomycorrhizal fungi. *Plant Soil* 172: 17-27.

BUCKLEY GP (1992). Ecology and Management of coppice woodland (Hardcover).

BUÉE M, VAIRELLES D, GARBAYE J (2005) Year-round monitoring of diversity and potential metabolic activity of the ectomycorrhizal community in a beech (Fagus silvatica) forest subjected to two thinning regimes. *Mycorrhiza* 15(4): 235-245.

BYRD KB, PARKER VT, VOGLER DR, CULLINGS KW (2000). The influence of clear-cutting on ectomycorrhizal fungus diversity in a lodgepole pine (*Pinus contorta*) stand, Yellowstone National Park, Wyoming, and Gallatin National Forest, Montana. *Canadian Journal of Botany* 78: 149-156.

CAUSIN R, MONTECCHIO L, MUTTO ACCORDI S (1996). Probability of ectomycorrhizal infection in a declining stand of common oak. *Annales des Sciences Forestieres* 53: 743-752.

CLINE ET, AMMIRATI JF, EDMONDS RL (2005). Does proximity to mature trees influence ectomycorrhizal fungus communities of Douglas-fir seedlings? *New Phytologist* 166: 993-1009.

CUTINI A (2001). New Management options in chestnug coppices: an evaluation on ecological bases. *Forest Ecology and Management* 141 (3): 165-174.

DAHLBERG A (2001).Community ecology of ectomycorrhizal fungi: an advancing interdisciplinary field. *New Phytologist* 150: 555-562.

DE ROMAN M, CLAVERIA V, DE MIGUEL AM (2005). A revision of the description of ectomycorrhiza published since 1961. *Mycological Research* 109 (10)1:1063-1104.

DURALL DM, JONES MD, WRIGHT EF, KROEGER P, COATES KD (1999). Species richness of ectomycorrhizal fungi in cutblocks of different sizes in the Interior Cedar-Hemlock forests of northwestern British Columbia: sporocarps and ectomycorrhizae. *Canadian Journal of Forest*

Research 29:1322-1332.

FELLNER R, CAISOLVÀ V (1994). Ecological aspects of mycorrhizae decline and oak dying in the Czech Republic. Estratti del convegno: "Environmental constraints and oaks: ecological and physiological aspects", Nancy (Francia) 29 agosto-1 settembre 1994.

GEBHARDT S, NEUBERT J, WÖLLECKE B, MÜNZENBERGER B, HÜTTL RF (2007). Ectomycorrhiza communities of red oak (*Quercus rubra* L.) of different age in the Lusatian lignite mining distict, East Germany. *Mycorrhiza* 17: 279-290.

GROGAN P, BAAR J, BRUNS TD (2000).Below-ground ectomycorrhizal community structure in a recently burned bishop pine forest. *Journal of Ecology* 88: 1051-1062.

HAGERMAN SM, JONES MD, BRADFIELD GE, GILLESPIE M, DURALL DM (1999). Effects of clear-cut logging on the diversity and persistence of ectomycorrhizae at a subalpine forest. *Canadian Journal of Forest Research* 29: 124-134.

HORTON TR, BRUNS TD (2001). The molecular revolution in ectomycorrhizal ecology: pecking into black-box. *Molecular Ecology* 10: 1855-1832.

JAKUCS E, KOVACS GM, AGERER R, ROMSICS C, ERÖS-HONTI Z (2005). Morphological-anatomical characterization and molecular identification of *Tomentella stuposa* ectomycorrhizae and related anatomotypes. *Mycorrhiza* 15: 247-258.

JANY JL, MARTIN F, GARBAYE J (2003). Respiration activity of ectomycorrhizas from *Cenococcum geophilum* and *Lactarius* sp. in relation to soil water potential in five beech forests. *Plant and Soil* 255: 487-494.

JONES MD, DURALL DM, CAIRNEY JWG (2003). Ectomycorrhizal fungal communities in young forest stands regenerating after clearcut logging. *New Phytologist* 157: 399-422.

KOIDE RT, XU B, SHARDA J, LEKBERG Y, OSTIGUY N (2005). Evidence of species interactions within an ectomycorrhizal fungal community. *New Phytologist* 165:305-316.

KÕLJALG U, DAHLBERG A, TAYLOR AFS, LARSSON E, HALLENBERG N, STENLID J, LARSSON KH, FRANSSON PM, KÅRÉN, JONSSON L (2000). Diversity and abundance of resupinate thelephoroid fungi as ectomycorrhizal symbionts in Swedish boreal forests. *Molecular Ecology* 9: 1985-1996.

KÕLJALG U, JAKUCS E, BOKA K, AGERER R (2001). Three ectomycorrhiza with cystidia formed by different *Tomentella* species as revealed by rDNA ITS sequences and anatomical characteristics. *Folia Cryptog. Estonia Facs.* 38: 27-39.

KÕLJALG U, TAMMI H, TIMONEN S, AGERER R, SEN R (2002). ITA rDNA sequence-based phylogenetic analysis of *Tomentellopsis* species from boreal and temperate forests, and the identification of pink-type ectomycorrhizas. *Mycological Progress* 1: 81-92.

KJØLLER R (2006). Disproortionate abundance between ectomycorrhizal root tips and their associated mycelia. *FEMS Mycrobiology Ecology* (58)2: 214-224.

KUYPER TW, LANDEWEERT R (2002). Vertical niche differentiation by hyphae of ectomycorrhizal fungi in soil. *New Phytologist* 156: 321–326

MAHMOOD S, FINLAY RD, ERLAND S (1999). Effects of repeated harvesting of forest residues on the ectomycorrhizal community in a Swedish spruce forest. *New Phytologist* 142: 577-585.

MONTECCHIO L, CAUSIN R, ROSSI S, MUTTO ACCORDI S (2004). Changes in ectomycorrhizal diversity in a declining *Quercus ilex* coastal forest. *Phytopathologia Mediterranea* 43: 26-34.

MOSCA E, MONTECCHIO L, SELLA L, GARBAYE J (2007). Short-term effect of removing tree competition on the ectomycorrhizal status of a declining pedunculate oak forest (*Quercus robur* L.). *Forest Ecology and Management* 244: 129-140.

NEVES MACHADO MH (1995). La mycorrhization contrôllée d'*Eucalyptus globulus* au Portugal et l'effet de la sècheresse sur la symbiose ectomycorhizienne chez cette essence. PhD thesis, Universitè de Nancy, France. 156 pp.

ORLANDER G, GEMMEL P, HUNT J (1990). Site Preparation: A Swedish Overview. FRDA Report 105, ISSN 0835-0752 1-61, pp. 1–61.

SCATTOLIN L, MONTECCHIO L, AGERER R (2008). The ectomycorrhizal community structure in high mountain Norway spruce stands. Trees. DOI 10.1007/s00468-007-0164-9.

STICHER L, MAUCH-MANI B, METRAUX JP(1997). Systemic acquired resistance. *Annual Review of. Phytopathology* 35: 235-270.

SUTTON RF (1993). Mounding site preparation: a review of European and North-American experience. New For. 7, 151–192.

TAYLOR AFS (2002). Fungal diversity in ectomycorrhizal communities: sampling effort and species detection. *Plant and Soil* 244: 19–28.

TOLJANDER JF, EBERHARDT U, TOLJANDER YK, PAUL LR, TAYLOR AFS (2006). Species composition of an ectomycorrhzial fungal community along a local nutrient gradient in a boreal forest. *New Phytologist* 170: 873-884.

VAN LOON LC, BAKKER PAHM, PIETERSE CM (1998). Systematic resistance induce by rhizosphere bacteria. *Annual review of Phytopatology* 36: 453-483.

WARGO PM (1988). Root vitality and mycorrhizal status of different health classed of red spruce trees. *Phytophatology* 78: 1533.

Abstract – The ectomycorrhizal community structure in Beech coppices of different age

The species composition of ectomycorrhizal (ECM) fungal communities can be strongly influenced by the sylvicultural practises, abiotic and biotic factors, which determine interactions among the species. In order to determine the influence of the coppicing on EM community, shoot age, bedrock types, exposure, slope, humus features, soil conditions, sampling points locations were taken into account as the most representative and influencing factors in these soil ecological dynamics. In summer 2005, 2006 and 2007, in 7 [2-48-years-old] Beech [*Fagus sylvatica* (L.) Karst.] coppices located in the Province of Trento (northern Italy), a monitoring on the the root tips was applied to compare these sites, and to give an additional instrument like a synthetic biological indicator for the traditional management strategies.

In the present study the results confirmed the ectomycorrhizal community structure investigated in 7 beech coppices of different age was typical with the occurrence of few abundant species and many others with significantly lower abundance. *Cenococcum geophilum* was the most frequently detected species in each site and in each sample date. Morphological, anatomical and molecular investigations revealed a total of 60 anatomotypes. Of these 35 were unknown on *Fagus sylvatica* up to now. The investigations on the community composition can be considered a great contribution to the biodiversity of the Beech forest, with four detailed species descriptions: *Fagirhiza byssoporioides*, *Fagirhiza entolomoides*, *Fagirhiza stellata* and *Hygrophorus penarius*. Additional investigations using stable isotopes were necessary to understand the parasitic attitude shown by this species in these coppices.

The investigation of the ECM community composition (species richness evenness, and dispersion, vitality and rate of mycorrhization) in relation to shoot age and to the main ecological factors revealed the absence of a real reaction to the coppicing, and the major importance of the slope or other ecological conditions to understand the species distribution.

An aggregation of the species was releaved, but the species features didn't show a clear correlation with the ecological stand conditions, concerning the spatial distribution and the soil horizons.

The results suggest that the coppice treatment in Beech, didn't have a significant effect on the EM community structure since 2 until 48 years from coppicing. Considering the stability of the EM community as a bioindicator of the ecosystem resilience, it can be supposed that a rational coppicing treatment could be a sustainable human activity, compatible with the ecosystem dynamics under these environmental conditions. Two more EM descriptions were performed: *Pseudotomentella humicola* on *Picea abies* and *Sistotrema muscicola* on *Castanea sativa*.

Riassunto – **La struttura della comunità ectomicorrizica in cedui di faggio di diversa età.**

La composizione e la struttura delle comunità ectomicorriziche (EM) possono essere fortemente influenzate dalle pratiche selvicolturali, che si aggiungono ad altri fattori abiotici e biotici, che determinano le interazioni tra le specie. Sono stati compiuti degli studi per determinare l'effetto della ceduazione sulla comunità EM, considerando principalmente l'età dei polloni (epoca dell'ultima ceduazione), tipo di substrato, l'esposizione, la pendenza dei siti, le caratteristiche delle forme di humus, del suolo e le condizioni di prelievo, perché meglio descrivevano la dinamica ecologia del suolo. E' stato realizzato un monitoraggio sfruttando un campionamento basato sullo studio degli apici radicali, effettuando diversi prelievi negli anni 2005, 2006 e 2007 in cedui di faggio[*Fagus sylvatica* (L.) Karst.] (ceduati 2-48 anni fa) situati nella provincia di Trento (Nord Italia). Lo scopo dello studio alla base di questo monitoraggio è stato di fornire uno strumento addizionale (come l'indice di stato ectomicorrizico) a quelli usati tradizionalmente per la gestione di questi siti. La struttura della comunità EM nei cedui oggetto di studio è apparsa riconducibile a quella tipicamente riportata in letteratura, con presenza di poche specie frequenti e una maggioranza di specie rare. *Cenococcum geophilum* è la specie dominante in tutti i campionamenti e in tutti i siti. Le analisi morfologiche, anatomiche e molecolari hanno permesso di definire 60 anatomotipi. Queste ricerche hanno contribuito allo studio della biodiversità nei boschi di faggio, con 35 specie mai osservate fino ad ora. Tra queste, 4 specie sono state descritte in dettaglio: *Fagirihiza byssoporioides*, *F. entolomoides*, *F. stellata* e *Hygrophorus penarius*.
Quest'ultima specie è stata oggetto di ulteriori indagini, impiegando anche saggi agli isotopi, perché ha manifestato un'attitudine parassitaria nel 2006 e nel 2007. Gli studi condotti sui principali parametri che permettono di definire la struttura di una comunità EM (come la ricchezza di specie, la dispersione delle stesse, la vitalità e il grado di micorrizazione) in relazione all'età dei polloni e ai fattori ecologici principali, ha permesso di definire un'assenza di reazione da parte della comunità stessa alla ceduazione. La distribuzione delle specie è maggiormente correlata ad alcune variabili stazionali (pendenza e umidità), che alla frequenza di ceduazione. E' stata rilevata un'aggregazione delle specie nei siti, ma nessuna chiara correlazione tra concerne le caratteristiche ecologiche delle stesse e le condizioni stazionali, sia per quanto riguarda la distribuzione superficiale sia per la distribuzione secondo il profilo umico. I risultati preliminari di questo studio suggeriscono l'ipotesi che la ceduazione in boschi di faggio con queste caratteristiche stazionali, non abbia alcun effetto significativo sulla struttura della comunità EM per un'età del ceduo compresa tra i 2 e i 48 anni. Se si considera inoltre la stabilità del consorzio EM come bioindicatore della resilienza dell'ecosistema, si può supporre che la razionale ceduazione possa costituire in questi siti, un esempio d'attività

selvicolturale sostenibile.

Grazie alle ricerche condotte in questi tre anni, è stato inoltre possibile realizzare altri due contributi, allo studio della diversità delle specie ectomicorriziche nelle foreste temperate e boreali, riportati in questa tesi: le descrizioni della specie *Pseudotomentella humicola* su *Picea abies* e della specie *Sistotrema muscicola* su *Castanea sativa*.

Acknowledgements

The PhD scholarship was supported by the Centro di Ecologia Alpina (TN, Italy), through the "Fondo per i progetti di ricerca della Provincia autonoma di Trento", "Inhumus Project InHumusNat2000" (1587/2004), and partially by the Foundation "Ing. A. Gini" (Università di Padova).

Un sincero ringraziamento ai miei genitori, a cui dedico questo lavoro, per la loro inesauribile pazienza e dedizione. Grazie anche ai compagni di viaggio, che mi hanno sostenuta, rendendo questa esperienza molto bella ed intensa.

"Man muss das Unmögliche versuchen, um das Mögliche zu erreichen" H. Hesse

Die VDM Verlagsservicegesellschaft sucht für wissenschaftliche Verlage abgeschlossene und herausragende

Dissertationen, Habilitationen, Diplomarbeiten, Master Theses, Magisterarbeiten usw.

für die kostenlose Publikation als Fachbuch.

Sie verfügen über eine Arbeit, die hohen inhaltlichen und formalen Ansprüchen genügt, und haben Interesse an einer honorarvergüteten Publikation?

Dann senden Sie bitte erste Informationen über sich und Ihre Arbeit per Email an *info@vdm-vsg.de*.

Sie erhalten kurzfristig unser Feedback!

VDM Verlagsservicegesellschaft mbH
Dudweiler Landstr. 99   Telefon +49 681 3720 174
D - 66123 Saarbrücken   Fax   +49 681 3720 1749
www.vdm-vsg.de

Die VDM Verlagsservicegesellschaft mbH vertritt

Printed by Books on Demand GmbH, Norderstedt / Germany